VOLUME ONE HUNDRED AND SIXTY ONE

PROGRESS IN MOLECULAR BIOLOGY AND TRANSLATIONAL SCIENCE

G Protein Signaling Pathways in Health and Disease

VOLUME ONE HUNDRED AND SIXTY ONE

PROGRESS IN MOLECULAR BIOLOGY AND TRANSLATIONAL SCIENCE

G Protein Signaling Pathways in Health and Disease

Edited by

YA-XIONG TAO

Department of Anatomy, Physiology, and Pharmacology, College of Veterinary Medicine;
Center for Neuroscience Initiative, Auburn University, Auburn, AL, United States

ELSEVIER

ACADEMIC PRESS
An imprint of Elsevier

Academic Press is an imprint of Elsevier
50 Hampshire Street, 5th Floor, Cambridge, MA 02139, United States
525 B Street, Suite 1650, San Diego, CA 92101, United States
The Boulevard, Langford Lane, Kidlington, Oxford OX5 1GB, United Kingdom
125 London Wall, London EC2Y 5AS, United Kingdom

First edition 2019

Notices
Knowledge and best practice in this field are constantly changing. As new research and experience broaden our understanding, changes in research methods, professional practices, or medical treatment may become necessary.

Practitioners and researchers must always rely on their own experience and knowledge in evaluating and using any information, methods, compounds, or experiments described herein. In using such information or methods they should be mindful of their own safety and the safety of others, including parties for whom they have a professional responsibility.

To the fullest extent of the law, neither the Publisher nor the authors, contributors, or editors, assume any liability for any injury and/or damage to persons or property as a matter of products liability, negligence or otherwise, or from any use or operation of any methods, products, instructions, or ideas contained in the material herein.

ISBN: 978-0-12-816724-3
ISSN: 1877-1173

For information on all Academic Press publications
visit our website at https://www.elsevier.com/books-and-journals

Publisher: Zoe Kruze
Acquisition Editor: Ashlie M. Jackman
Editorial Project Manager: Joanna Collett
Production Project Manager: James Selvam
Cover Designer: Matthew Limbert

Typeset by SPi Global, India

Contents

Contributors

Dan-Na Chen
Department of Basic Medical Sciences, Changsha Medical University, Changsha, China

Dong-Yu Guo
Department of Clinical Laboratory, Xiamen Huli Guoyu Clinic, Co., Ltd., Xiamen, China

Eugenia V. Gurevich
Department of Pharmacology, Vanderbilt University, Nashville, TN, United States

Vsevolod V. Gurevich
Department of Pharmacology, Vanderbilt University, Nashville, TN, United States

Bing Han
Department of Endocrinology, Shanghai Ninth People's Hospital, Shanghai Jiao Tong University School of Medicine, Shanghai, People's Republic of China

Zhi-Shuai Hou
Department of Anatomy, Physiology and Pharmacology, College of Veterinary Medicine, Auburn University, Auburn, AL, United States

Hong Jia
School of Life Sciences; Hunan Key Laboratory of Animal Models for Human Diseases; Hunan Key Laboratory of Medical Genetics, Central South University, Changsha, China

Jing Kou
International Education College, Henan University, Kaifeng, China

Yi-Qun Kuang
Center for Translational Medicine, Huaihe Clinical College, Huaihe Hospital of Henan University, Kaifeng, China

Jia-Da Li
School of Life Sciences; Hunan Key Laboratory of Animal Models for Human Diseases; Hunan Key Laboratory of Medical Genetics, Central South University, Changsha, China

Jie Qiao
Department of Endocrinology, Shanghai Ninth People's Hospital, Shanghai Jiao Tong University School of Medicine, Shanghai, People's Republic of China

Wenmin Sun
State Key Laboratory of Ophthalmology, Zhongshan Ophthalmic Center, Sun Yat-sen University, Guangzhou, China

Ya-Xiong Tao
Department of Anatomy, Physiology, and Pharmacology, College of Veterinary Medicine; Center for Neuroscience Initiative, Auburn University, Auburn, AL, United States

Wei Wang
Department of Clinical Laboratory, Xiamen Huli Guoyu Clinic, Co., Ltd., Xiamen, China

Xinying Wang
School of Life Sciences; Hunan Key Laboratory of Animal Models for Human Diseases; Hunan Key Laboratory of Medical Genetics, Central South University, Changsha, China

Jiayu Wu
School of Life Sciences; Hunan Key Laboratory of Animal Models for Human Diseases; Hunan Key Laboratory of Medical Genetics, Central South University, Changsha, China

Li-Kun Yang
Department of Anatomy, Physiology and Pharmacology, College of Veterinary Medicine, Auburn University, Auburn, AL, United States

Qingjiong Zhang
State Key Laboratory of Ophthalmology, Zhongshan Ophthalmic Center, Sun Yat-sen University, Guangzhou, China

Zheng-Rui Zhang
Department of Anatomy, Physiology and Pharmacology, College of Veterinary Medicine, Auburn University, Auburn, AL, United States; Key Laboratory of Marine Genetics and Breeding, Ministry of Education, College of Marine Life Sciences, Ocean University of China, Qingdao, China

Yaguang Zhao
School of Life Sciences; Hunan Key Laboratory of Animal Models for Human Diseases; Hunan Key Laboratory of Medical Genetics, Central South University, Changsha, China

Preface

G protein-coupled receptors (GPCRs), also called seven-transmembrane domain receptors, are wonderful molecular machines that sit on the plasma membrane, transmitting diverse extracellular signals to the inside of the cells to maintain homeostasis in the organism. About 800 GPCR genes exist in the human genome. They participate in the regulation of almost all physiological processes. Hence, these receptors have been actively pursued as therapeutic targets. Indeed, these pursuits have been handsomely rewarded, with close to 40% of current drugs targeting (only a small portion) of the GPCRs, including some of the best-selling drugs.

Since 1990, there have been extensive studies on mutations in genes encoding these GPCRs and human diseases. Two volumes on this topic were published in 2009 (Volumes 88 and 89) in *Progress in Molecular Biology and Translational Science*. Dr. P. Michael Conn was the Serial Editor at that time. He was guiding me every step of the way when I was working on these two volumes. Unfortunately, he died suddenly around Thanksgiving holiday in 2016. I dedicate this volume to his memory.

The current volume is a continuation, covering several new GPCRs, as well as other molecules in G protein-signaling pathway, such as arrestins and cyclic nucleotide-gated channel. Again, the focus is mutations and polymorphisms and human illness. Whenever, potential therapeutic strategies for correcting defects in these mutations are highlighted. The last chapter is specifically directed to this aspect, illustrating the potential strategies for correcting nonsense mutations, mutations that result in mutant receptors retained intracellularly by cellular quality control, and mutations that disrupt receptor binding to the ligand or coupling/activation of the heterotrimeric G proteins. It is hoped that future studies along these lines can be translated into precision medicine for patients harboring these mutations.

I thank all the authors for turning in their excellent chapters on time, and going over several revisions as requested. The encouragement and support from Dr. David B. Teplow, Serial Editor, is greatly appreciated. I also thank the colleagues at Elsevier for their help in various stages of this project, including Ashlie Jackman, Joanna M. Collett, Indhumathi Mani, and Naveen Kari. It has been a wonderful experience to see this book from conception to print.

Ya-Xiong Tao
Auburn, AL, USA

CHAPTER ONE

Diseases associated with mutations in *CNGA3*: Genotype–phenotype correlation and diagnostic guideline

Wenmin Sun, Qingjiong Zhang*
State Key Laboratory of Ophthalmology, Zhongshan Ophthalmic Center, Sun Yat-sen University, Guangzhou, China
*Corresponding author: e-mail addresses: zhangqji@mail.sysu.edu.cn; zhangqingjiong@gzzoc.com

Contents

Abstract

Along with the molecular and functional characterization of *CNGA3*, knowledge about diseases associated with *CNGA3* mutations has made great progress. So far, *CNGA3* mutations are not only one of the most common causes of achromatopsia and cone dystrophy or cone-rod dystrophy but also one of the most commonly mutated genes among various forms of retinopathy. Understanding the clinical characteristics of

Progress in Molecular Biology and Translational Science, Volume 161
ISSN 1877-1173
https://doi.org/10.1016/bs.pmbts.2018.10.002

CNGA3-associated retinal diseases may help clinical practice of infants or children with related diseases. Recognizing the importance of *CNGA3* in inherited retinal diseases may enhance related research in searching for functional restoration or repair of CNGA3 defects.

Abbreviations

ACHM	achromatopsia
cAMP	cyclic adenosine monophosphate
cGMP	cyclic guanosine monophosphate
CNG	cyclic nucleotide-gated
CORD	cone dystrophy or cone-rod dystrophy
dbSNP	the single nucleotide polymorphism database
Ensembl	Ensembl project
ERG	electroretinogram
ExAC	the exome aggregation consortium
GRCh	the genome reference consortium human genome build
HGNC	HUGO Gene Nomenclature Committee
LCA	Leber congenital amaurosis
OCT	Oligocone trichromacy
OMIM	Online Mendelian Inheritance in Man
RefSeq	the reference sequence
RetNet	retinal information network
UniProtKB	the UniProt knowledgebase

1. *CNGA3* overview

CNGA3 is the currently approved symbol of a gene called *cyclic nucleotide-gated channel alpha 3*, with different identification numbers at different systems (HGNC: 2150; Entrez Gene: 1261; Ensembl: ENSG00000144191; OMIM: 600053; UniProtKB: Q16281). Previously, this gene was also named as *CNG3*, *ACHM2*, *CCNC1*, *CCNCa*, *CNCG3*, or *CCNCalpha*. It is located in 2q11.2 at genomic 98346155..98399154 forward strand based on GRCh38.p12 with eight exons (seven coding exons). This gene was cloned in 1994 by polymerase chain reaction using two primers derived from two conserved regions in *CNGA1* and *CNGA2*.[1] Two alternatively spliced transcripts have been described based on RefSeq although two additional transcripts are predicted (https://www.ncbi.nlm.nih.gov/nuccore/NC_000002.12?from=98346155&to=98399154&report=genbank).

CNGA3 encodes an alpha subunit of the cyclic nucleotide-gated cation channel protein, a key molecule associated with visual signal transduction. Cyclic nucleotide-gated cation channels or CNG channels are directly activated through binding of cyclic nucleotides, including cGMP and cAMP, and play important roles in retinal photoreceptors and olfactory receptor neurons.[2] A CNG channel consists of four subunits divided into two subtypes, CNGA and CNGB.[2] In vertebrates, the CNG channels are encoded by six genes, including four for CNGA (*CNGA1*, *CNGA2*, *CNGA3*, *CNGA4*) and two for CNGB (*CNGB1* and *CNGB3*). The human rod photoreceptor CNG channels contain two subunits (CNGA1 and CNGB1) that form heterotetramer,[3,4] consisting of three A1 subunits and one B1 subunit.[5] Biallelic mutations in either *CNGA1* or *CNGB1* result in autosomal recessive retinitis pigmentosa.[6,7] Likewise, the human cone photoreceptor CNG channels also contain two subunits (CNGA3 and CNGB3) that form heterotetramer,[8–10] consisting of two A3 subunits and two B3 units.[9] Biallelic mutations in either *CNGA3* or *CNGB3* cause autosomal recessive achromatopsia[11,12] and several related diseases. Mutations in the genes resulted in loss of functional cone CNG channels or impaired trafficking or function of the channels.[13] In addition to *CNGA3* and *CNGB3*, mutations in four other genes, including *GNAT2*, *PDE6C*, *PDE6H*, and *ATF6*, are reported to cause achromatopsia (Table 1), all as autosomal recessive trait.[14–17]

2. Diseases associated with mutations in *CNGA3*

The association of *CNGA3* with human diseases started from a genome-wide linkage analysis on an inbred Jewish family from Iran, in which achromatopsia (ACHM) in the family was mapped to chromosome 2 in 1997.[18] The linkage interval was refined to a 3 cM region harboring the *CNGA3* gene based on the linkage analysis of eight families with rod monochromacy, an autosomal recessive disease with total color blindness.[19] Subsequent sequencing analysis identified biallelic mutations of *CNGA3* in five families with rod monochromacy previously studied by linkage analysis. Since then, a number of mutations in *CNGA3* have been detected in patients with different forms of autosomal recessively inherited cone dominated retinal diseases, based on the phenotype-guided (or diagnosis-guided) genotype analysis: (1) achromatopsia 2 (ACHM2), including complete ACHM (also called rod monochromacy or total color blindness) and some cases of incomplete ACHM; (2) cone dystrophy or cone-rod dystrophy (CORD)[20]; (3) Leber congenital amaurosis (LCA)[21]; and (4) Oligocone trichromacy (OCT).[22]

Table 1 Clinical characteristics of different forms of *CNGA3*-associated retinopathy

Classification	ACHM		OCT	CORD	LCA
	Complete	Incomplete			
Signs and symptoms					
Congenital nystagmus	+	+	+/−	+/−	+[a]
Photophobia	++	++	+	+/−	+/−
Reduced visual acuity[b]	+++	++ ~ +++	+ ~ ++	+ ~ +++	++++
Colorblindness[b,c]	++++	+++	−	+/−	+/−
Fundus changes in infant[d]					
Posterior	−	−	−	+/−	+/−
Mid-peripheral	−	−	−	+/−	+
Electroretinogram					
Rod response	N	N	N	MR ~ SR	NR ~ SR
Cone response at light	NR	SR	NR ~ SR	MR ~ NR	NR ~ SR
Systemic features	−	−	−	+/−	+/−
Other genes for this disease[e]	5	6	6	34	25
	CNGA3	*CNGA3*	*CNGA3*	*CNGA3*	*CNGA3*
	CNGB3	*CNGB3*	*CNGB3*	*CNGB3*	*CEP290*
	GNAT2	*GNAT2*	*GNAT2*	*GNAT2*	Plus 23
	PDE6C	*PDE6C*	*CEP290*	*PDE6C*	
	ATF6	*ATF6*	*RGS9*	*PDE6H*	
		PDE6H	*R9AP*	Plus 29	

[a]Mostly roving nystagmus but occasionally pendular nystagmus.
[b]These two signs are hardly measurable in infant.
[c]Mostly total colorblindness in ACHM but partial color anomaly in CORD.
[d]Fundus changes may become obvious in CORD and LCA with age.
[e]The other five genes for ACHM: *CNGB3*, *GNAT2*, *PDE6C*, *PDE6H*, and *ATF*; for OCT: *CNGB3*, *GNAT2*, *CEP290*, *RGS9*, *R9AP*; for CORD and LCA: referring the whole list in RetNet (https://sph.uth.edu/retnet/sum-dis.htm).
+, present; −, absent; *N*, normal; *NR*, nonrecordable; *MR*, mildly reduced; *SR*, severely reduced.

Currently, at least 144 mutations in *CNGA3* have been identified in 340 families, including 264 families with ACHM, 48 families with CORD, 3 families with OCT, 2 families with LCA, and 23 families with unclassified retinal dystrophies (Table 2).[11,13,20–69] Most of this information is based on the mutation detection in a cohort of patients with one or a few forms of retinal diseases, i.e., phenotype-guided (or diagnosis-guided) genotype analysis. The proportion of ACHM may be overrepresented since most of the *CNGA3* mutations are detected based on patients with ACHM but rarely on patients with other retinal diseases.

Table 2 *CNGA3* mutations and their associated phenotypes reported so far

Genotype	**Number of families reported**					
(Amino acid)	**Total**	**ACHM**	**CORD**	**OCT**	**LCA**	**UN**
p.[T14Pfs*]; [R283W]	1	1				
p.[S21*]; [S21*]	1	1				
p.[S21*]; [R410W]	1	1				
p.[S21*]; [p.V529M]	1	1				
p.[S21*]; [E590K]	2	1	1			
p.[S21*]; [L626R]	1	1				
p.[R23*]; [R23*]	3	3				
p.[R23*]; [D260N]	1		1			
p.[R23*]; [E344*]	1	1				
p.[R23*]; [R427C]	1		1			
p.[R23*]; [F547 L]	1	1				
p.[R23*]; [R563H]	1		1			
p.[H36fs]; [K512 M]	1					1
p.[H36fs]; [R563H]	1					1
p.[A51fs*]; [A51fs*]	3	3				
p.[Q47*]; [R499*]	1	1				
p.[G49 fs*]; [W316*]	1					1
p.[W83*]; [F644S]	1		1			
c.[396-11C>G]; [396-11C>G]	1		1			
c.[396-11C>G]; [1074G>A]	3	2	1			

Continued

Table 2 *CNGA3* mutations and their associated phenotypes reported so far—cont'd

Genotype	Number of families reported					
(Amino acid)	Total	ACHM	CORD	OCT	LCA	UN
c.[396-11C>G]; [1495C>T]	1		1			
c.[396-11C>G]; [1556 T>C]	1		1			
c.[396-11C>G]; [1627_1635del]	1		1			
p.[K155Rfs*]; [K155Rfs*]	1	1				
p. [V160G]; [F547L]	1	1				
p.[D162V]; [D260N]	3	2				1
p.[D162V]; [Q604*]	1		1			
p.[P163L]; [P163L]	1	1				
p.[P163L]; [R283W]	1	1				
p.[P163L]; [R436W]	1	1				
p.[W171*]; [W171*]	1		1			
p.[W171*]; [E590K]	1		1			
p.[W171C]; [R277H]	1	1				
p.[W171C]; [R221*]	1	1				
p.[W171C]; [R283W]	1		1			
p.[V179D]; [V179D]	1	1				
p.[Y181C]; [Y181C]	2	2				
p.[N182Y]; [F547L]	2	2				
p.[L186F]; [L186F]	2	1				1
p.[L186F]; [R277C]	1	1				
p.[I187T]; [I187T]	1		1			
p.[C191Y]; [C191Y]	1	1				
p.[C191Y]; [R277C]	1	1				
p.[E194K]; [F547L]	1	1				
p.[Q196*]; [Q196*]	1	1				
p.[Q196*]; [L380del]	1	1				
p.[D211E]; [D211E]	1	1				
p.[D211E]; [V336F]	2	2				
p.[R221*]; [R221*]	2	1	1			

Table 2 *CNGA3* mutations and their associated phenotypes reported so far—cont'd

Genotype	**Number of families reported**					
(Amino acid)	**Total**	**ACHM**	**CORD**	**OCT**	**LCA**	**UN**
p.[R221*]; [R277C]	2	1	1			
p.[R221*]; [R277H]	1	1				
p.[R221*]; [R427C]	1	1				
p.[R221*]; [V529 M]	1		1			
p.[A222P]; [G557R]	1	1				
p.[R223G]; [G557R]	1	1				
p.[R223Q]; [R223W]	1	1				
p.[R223Q]; [R436W]	1	1				
p.[R223W]; [R223W]	3	1	1			1
p.[R223W]; [R277H]	1	1				
p.[R223W]; [R283Q]	1	1				
p.[R223W]; [P372S]	1	1				
p.[R223W]; [S341P]	2	2				
p.[R223W]; [R427C]	4	2	1			1
p.[R223W]; [R410W]	1	1				
p.[R223W]; [V451 fs*]	1	1				
p.[R223W]; [F547L]	1	1				
p.[R223W]; [E590K]	1	1				
p.[T224R]; [T369S]	2	2				
p.[T224I]; [R283W]	1		1			
c.[674-2A>C]; [671C>T]	1	1				
p.[E228K]; [E228K]	1	1				
p.[F249S]; [E344*]	1	1				
p.[D252N]; [D252N]	1					1
p.[P258R]; [V373Rfs*]	1		1			
p.[D260N]; [F547L]	1	1				
p.[D260N]; [G548R]	1	1				
p.[D260N]; [R277G]	1	1				
p.[A262P]; [F547L]	1	1				

Continued

Table 2 *CNGA3* mutations and their associated phenotypes reported so far—cont'd

Genotype	**Number of families reported**					
(Amino acid)	**Total**	**ACHM**	**CORD**	**OCT**	**LCA**	**UN**
p.[Y263D]; [T565M]	1	1				
p.[G267D]; [Q655*]	3	3				
p.[P271A]; [G329C]	1	1				
p.[R274K]; [L278P]	1	1				
p.[R274S]; [R274S]	2	2				
p.[N276S]; [N276S]	1	1				
p.[R277C]; [R277C]	3	3				
p.[R277C]; [R277G]	1	1				
p.[R277C]; [R283W]	3	1	2			
p.[R277C]; [R410W]	1	1				
p.[R277C]; [R427C]	1	1				
p.[R277C]; [R436W]	1					1
p.[R277C]; [R439W]	1	1				
p.[R277C]; [A469T]	1			1		
p.[R277C]; [L527R]	1	1				
p.[R277C]; [Q537*]	1					1
p.[R277C]; [F547 L]	4	4				
p.[R277C]; [G557R]	1	1				
p.[R277C]; [R563C]	2	2				
p.[R277C]; [Y573C]	2	1				1
p.[R277G]; [G329C]	1	1				
p.[R277G]; [S341P]	1	1				
p.[R277G]; [G557R]	1	1				
p.[R277H]; [R277H]	2	2				
p.[R277H]; [R283Q]	1		1			
p.[R277H]; [Y357C]	1	1				
p.[R277H]; [W358*]	1		1			
p.[R277H]; [E376del]	1	1				
p.[R277H]; [R439W]	1	1				

Table 2 *CNGA3* mutations and their associated phenotypes reported so far—cont'd

Genotype	**Number of families reported**					
(Amino acid)	**Total**	**ACHM**	**CORD**	**OCT**	**LCA**	**UN**
p.[R277H]; [V529M]	1		1			
p.[R277H]; [R569H]	1	1				
p.[R277H]; [G557R]	1	1				
p.[L278P]; [R436W]	1	1				
p.[R283W]; [R283W]	17	14				3
p.[R283W]; [P372S]	1	1				
p.[R283W]; [F380S]	1	1				
p.[R283W]; [G397V]	1	1				
p.[R283W]; [R427C]	1	1				
p.[R283W]; [R436W]	1	1				
p.[R283W]; [R499*]	1	1				
p.[R283W]; [F547L]	4	4				
p.[R283W]; [R563H]	1	1				
p.[R283W]; [T565M]	1	1				
p.[R283W]; [S570I]	1	1				
p.[R283Q]; [R283Q]	2	1	1			
p.[R283Q]; [R427C]	1	1				
p.[R283Q]; [L433W]	2	2				
p.[R283Q]; [R436W]	1	1				
p.[R283Q]; [G557R]	2	1				1
p.[R283Q]; [R569H]	1	1				
p.[T291R]; [F547L]	2	2				
p.[T291Rfs*]; [V529 M]	1		1			
p.[T291Rfs*]; [K659*]	1	1				
p.[R302G]; [F547C]	1	1				
p.[I314del]; [I314del]	11	11				
p.[I314del]; [R427C]	1	1				
p.[I314del]; [D485V]	3	3				
p.[I314del]; [V529M]	2	2				

Continued

Table 2 *CNGA3* mutations and their associated phenotypes reported so far—cont'd

Genotype	**Number of families reported**					
(Amino acid)	**Total**	**ACHM**	**CORD**	**OCT**	**LCA**	**UN**
p.[I314del]; [F547C]	1	1				
p.[I314del]; [R661S]	1	1				
p.[A318T]; [A318T]	1					1
p.[C319R]; [C319R]	2		2			
p.[C319R]; [G525S]	1				1	
p.[C319fs*]; [F547L]	1	1				
p.[F322S;R436Q]; [W440*]	1	1				
p.[A323D]; [A323D]	1	1				
p.[A323P]; [A469T]	1		1			
p.[G329C]; [G329C]	1	1				
p.[G329C]; [R436W]	2	2				
p.[F330S]; [R439W]	1		1			
p.[F330S]; [V529 M]	1		1			
p.[G331R]; [G331R]	1					1
p.[D333Lfs*]; [M424V]	1	1				
p.[S334F]; [R439W]	1		1			
p.[Y337Sfs*]; [D543_S545del]	1	1				
p.[E344*]; [R436W]	1	1				
p.[I353T]; [E377del]	1	1				
p.[Y357C]; [T565M]	2	1		1		
p.[W358*]; [R569H]	1	1				
p.[W358*]; [R439W]	1	1				
p.[W358*]; [E590K]	1	1				
p.[W358*]; [V529M]	1		1			
p.[L363P]; [L363P]	1	1				
p.[G367V]; [G367V]	3	3				
p.[P372S]; [P372S]	5	4				1
p.[P372S]; [R661S]	1	1				

Table 2 *CNGA3* mutations and their associated phenotypes reported so far—cont'd

Genotype	**Number of families reported**					
(Amino acid)	**Total**	**ACHM**	**CORD**	**OCT**	**LCA**	**UN**
p.[V373*]; [L527M]	1	1				
p.[E376K]; [E376K]	5	5				
p.[V383Rfs*]; [V529M]	1	1				
p.[S401P]; [S401P]	1	1				
p.[S401P]; [R427C]	1	1				
p.[R410W]; [R410W]	1					1
p.[R410W]; [D514V]	1	1				
p.[R410W]; [V529 M]	1	1				
p.[R427C]; [R436W]	1	1				
p.[R427C]; [F547L]	1	1				
p.[R427C]; [V451fs]	2	2				
p.[R427C]; [L527R]	1	1				
p.[R427C]; [M519I]	1		1			
p.[D432fs*]; [V529M]	1	1				
p.[L433W]; [unknown]	1					1
p.[R436W]; [R436W]	3	2	1			
p.[R436W]; [W440Cfs*]	1	1				
p.[R436W]; [R499*]	1		1			
p.[R436W]; [V529M]	1	1				
p.[R436W]; [F547L]	1	1				
p.[R436W]; [R563H]	1	1				
p.[R436W]; [L633P]	1	1				
p.[R439W]; [D543_S545del]	1	1				
p.[W440*]; [G516E]	1	1				
p.[V456fs*; G557R]	1			1		
p.[R463W]; [R463W]	1					1
p.[L464R]; [L541F]	1	1				
p.[L464R]; [F547L]	1	1				

Continued

Table 2 *CNGA3* mutations and their associated phenotypes reported so far—cont'd

Genotype	**Number of families reported**					
(Amino acid)	**Total**	**ACHM**	**CORD**	**OCT**	**LCA**	**UN**
p.[I482Hfs*]; [G548R]	1	1				
p.[I482Hfs*]; [R569H]	2	2				
p.[C486Y]; [C486Y]	1					1
p.[R499*]; [V529M]	1	1				
p.[R499*]; [unknown]	1	1				
p.[P505A]; [G513R]	1		1			
p.[C510S]; [C510S]	2	2				
p.[G513E]; [R569H]	2	2				
p.[I522T]; [F547L]	2	1				1
p.[G525D]; [T565M]	1	1				
p.[L527M]; [L527M]	1				1	
p.[L527R]; [L527R]	1	1				
p.[V529M]; [V529M]	15	13	2			
p.[V529M]; [E590K]	1	1				
p.[V529M]; [G548R]	1	1				
p.[V529M]; [S570I]	1		1			
p.[D533H]; [D533H]	1	1				
p.[V540I]; [V540I]	1		1			
p.[F547L]; [F547L]	17	16	1			
p.[F547L]; [G561E]	1	1				
p.[F547L]; [R563H]	2		2			
p.[G548R]; [G548R]	1	1				
p.[G557R]; [G557R]	2	2				
p.[S570I]; [S570I]	1	1				
p.[G684R]; [G684R]	1		1			
Total	340	264	48	3	2	23

*Stop codon.
UN, phenotype unclassified.

So far, it is unclear if additional forms of retinal diseases are caused by mutations in *CNGA3* since the genomic sequence of *CNGA3* has not been widely studied in patients with other forms of retinal dystrophies. Systemic sequencing evaluation of *CNGA3* in all forms of retinal diseases may reveal this point. Such genotype-guided phenotype clarification can be partly achieved by authenticating variants of *CNGA3* based on data derived from whole-exome sequencing on patients with different forms of retinal dystrophy. Analysis of our in-house data from at least 5000 probands (whole-exome sequencing, targeting exome sequencing, or Sanger sequencing) with different forms of genetic eye diseases detected biallelic mutations of *CNGA3* in a number of probands with CORD, ACHM, or LCA, in which CORD is the most common manifestation (Fig. 1).

Among the six genes responsible for ACHM, mutations in *CNGA3* are the second most common cause of ACHM in Caucasians, occurring in about 22% of families with ACHM and other related diseases (Fig. 2). However, *CNGA3* mutations are the most common cause of ACHM and CORD in Chinese,[47] accounting for 97% (34/35)[30,45,47,48] of ACHM with identified mutation in the six genes and over half of CORD with identified mutations in the 35 known CORD genes listed in RetNet (https://sph.uth.edu/retnet/sum-dis.htm).

For some genetic eye diseases, typical phenotypes are usually described in textbooks, but many similar but variant phenotypes are seen in individual patients in clinical practice. An infant with profound visual problem, such as photophobia and nystagmus, may be suspected to have ACHM2, CORD,

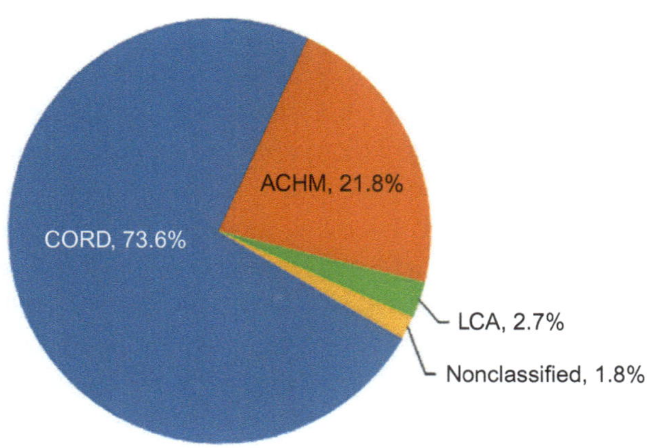

Fig. 1 Proportion of phenotypes associated with *CNGA3* mutations in Chinese based on whole-exome sequencing, targeted exome sequencing, or Sanger sequencing. For all probands with biallelic *CNGA3* mutations, CORD is the most common phenotype (73.6%), while ACHM and LCA account for 21.8% and 2.7%, respectively.

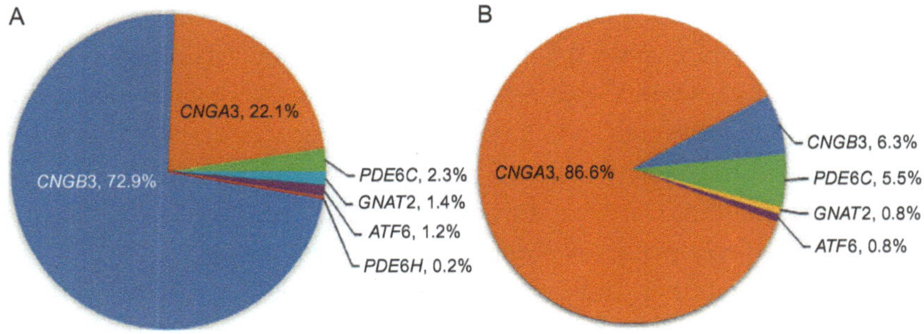

Fig. 2 Proportion of mutations detected in all the six genes disregarding phenotypes. (A) Data from Caucasians: *CNGB3* is the most commonly mutated gene, accounting for 72.9% of families, while *CNGA3* mutations are identified in 22.1% of families. (B) Data from Chinese: mutations in *CNGA3* accounts for 86.6%, while mutations in *CNGB3* only accounts for 6.3% families.

LCA, OCT, albinism, aniridia, etc., before undergoing thorough objective and subjective phenotypic assessment or prior to the full development of the ocular or systemic signs. Valuable signs and symptoms for differential diagnosis may be absent or too subtle to be identified in infant. These problems are also present in different forms of *CNGA3*-associated retinopathy as detailed in Table 1.

2.1 ACHM caused by *CNGA3* mutations

CNGA3 biallelic mutations are the most common cause of ACHM in Chinese[47,48] but are the second most common cause of ACHM in Caucasians[39,61] (Fig. 2). Both complete and incomplete ACHM can be caused by biallelic mutations in *CNGA3*. Patients with complete ACHM are characterized by severe photophobia, rapid pendular nystagmus, greatly reduced visual acuity but better vision at dark or dim light, complete inability to discriminate color, normal fundus, and absent cone responses but normal rod responses on ERG recordings. Likewise, patients with incomplete ACHM have a similar phenotype as those with complete ACHM but generally less severe, such as slight better visual acuity and severely reduced cone responses but normal rod responses on ERG recordings. For a child with nystagmus and photophobia but who cannot tell their visual impairment, it is usually hard to differentiate ACHM from other related diseases even after ocular examination and ERG test. This situation is common at initial visit of every affected child although follow-up examination for a few years can tell the difference (Table 1).

Infants with *CNGA3*-associated ACHM usually demonstrated profound photophobia and rapid horizontal pendular nystagmus noticed in the first few months after birth. Reduced visual acuity may be recognized subsequently. Parents usually noticed that the vision of their affected children improved obviously at night or under dim light. Total color blindness is usually not the presenting sign because infants cannot tell their problem or inability to discriminate between colors. Total color blindness may not be noticed even for children of several years old as they may mimic to differentiate colors by recognizing different brightness. Therefore, complete color blindness is usually revealed in children who can perform color vision test. Routine ophthalmic examination demonstrates normal cornea, lens, and fundus (Fig. 3). It is important to check the mid-peripheral fundus as well as posterior fundus since infants with other forms of retinopathy may exhibit normal-like posterior fundus at early stage but recognizable fundus changes at mid-peripheral retina. This may be neglected in routine clinical practice, in which posterior region of the fundus is usually visualized by routine ophthalmoscopic examination, especially in outpatient clinic with huge number of patients. Recognizable but subtle changes on mid-peripheral retina in infants may become invisible or less obvious on wide-field fundus images. Electroretinography is an important examination in differentiating ACHM from other diseases, showing normal rod photoreceptor responses but nonrecordable cone photoreceptor responses in typical form of ACHM. However, recent studies on ACHM have shown that the rod function may be slightly impaired in such patients and it may progress with age.[70] On the other hand, some patients may have residual cone function. For those patients with partly impaired rod function or residual cone function, they are more easily classified as CORD rather than ACHM based on clinical phenotypes, even after detection of mutations in *CNGA3*.

2.2 CORD caused by *CNGA3* mutations

CORD refers to a series of inherited retinal disorders with earlier or predominantly cone impairment and concurrent or subsequent less severe rod impairment. CORD is usually hard to differentiate with cone dystrophy, whereas some cone dystrophy may actually be the early stage of CORD and may subsequently develop to CORD with aging. Patients with CORD usually have reduced visual acuity, photophobia, color vision defects, central scotomata, with or without nystagmus, and absent or severely reduced cone responses accompanied with less severe rod impairment on ERG

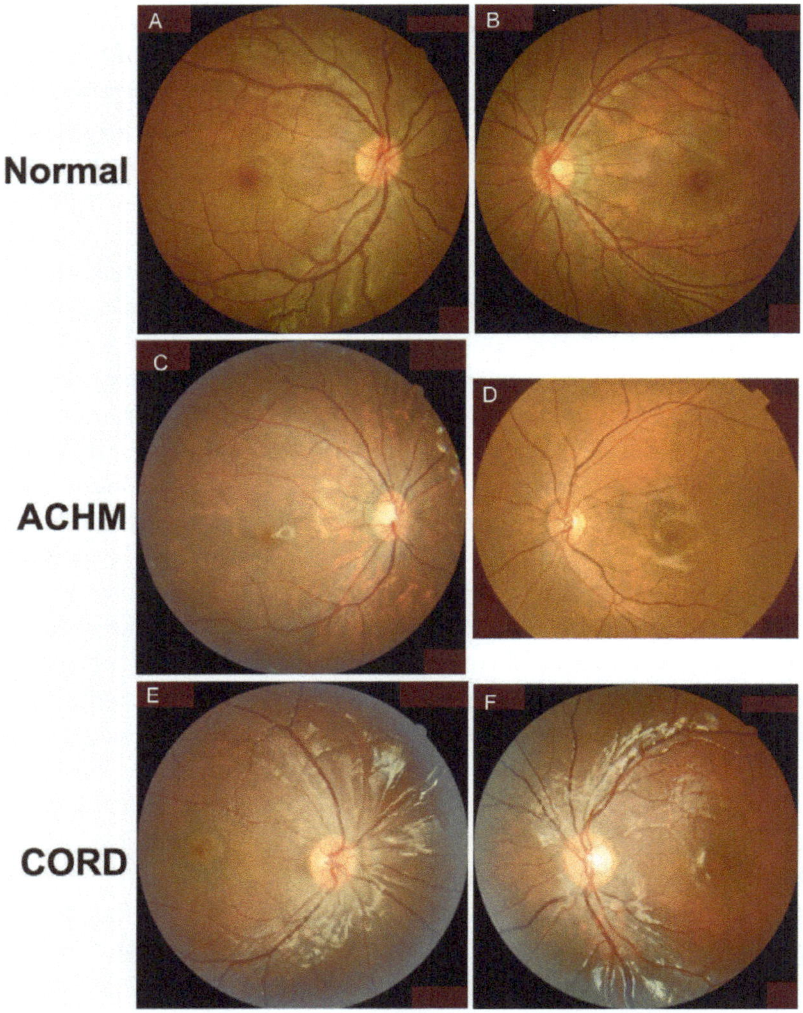

Fig. 3 Fundus photos of *CNGA3*-associated ACHM and CORD. (A and B) Fundus photos of left and right eyes from normal individuals with normal visual acuity. (C and D) Normal optic disc, posterior and mid-peripheral retina with normal foveal reflex were present in two patients with ACHM, (C) at age of 11 years old with c.[830G>A; 1315C>T] mutation and (D) at age of 7 years old with c.[396-11C>G;1074G>A] mutation. (E and F) Degenerative change with *yellowish dots* and loss of foveal reflex were observed in macular region in two patients with CORD, (E) with c.[778G>A;67C>T] mutation and at age of 5 years old and (F) with c.[458G>A;458G>A] mutation at age of 4 years old.

recordings. Fundus changes may vary from normal looking to fine macular lesion, macular atrophy, macular pigmentation, narrowing arteries, and optic disc pallor. At late stage, CORD may progress to generalized retinal degeneration involving both posterior and peripheral fundus, which might be indistinguishable from retinitis pigmentosa.

Biallelic mutations in *CNGA3* are the most common cause of CORD in Chinese,[47] accounting for over half of CORD with identified causative mutations.[38] CORD caused by *CNGA3* mutations was also found in other populations.[57] CORD has also been reported to associate with mutations in *CNGB3*,[60] in which mutations are the most common cause of ACHM in Caucasians.[15,71]

Infants with CORD caused by *CNGA3* mutations usually exhibit significant photophobia and nystagmus, which are noticed in first few months after birth. Ophthalmoscopic examination reveals a normal fundus or mild macular degeneration. ERG recordings demonstrated absent or severely reduced cone responses accompanied with mildly or moderately reduced rod responses. When they grew up to a few years old, they show better vision at dark or under dim light and their nystagmus may be slightly relieved. Fundus may be normal for many years in some cases, while it can progress to typical macular degeneration (Fig. 3).

For *CNGA3*-associated retinopathy, it might be problematic to differentiate CORD from ACHM by using rod values on ERG alone, as ACHM may have smaller than normal waveforms when compared to standard control values.[72] However, some patients do have significantly reduced rod responses compared to normal controls in clinic.[47] Long-term follow-up may tell stable ACHM from progressive CORD, in which the latter have progressive macular degeneration and progressive central scotomata. In clinic, a few patients with *CNGA3* mutations not only have ERG changes typical for cone-rod dystrophy but also had obvious macular degenerative change in early childhood (Fig. 3). The reasons why *CNGA3* mutations cause both stable and progressive retinopathy are currently unknown. However, several studies do observe progressive cone degeneration in patients with presumed stable ACHM caused by *CNGA3* mutations.[62,70,73] Progressive cone loss accompanied with progressive rod impairment, functionally and structurally, has been observed in mice with homozygous *CNGA3* mutations.[74,75] These lines of evidence suggest that *CNGA3* mutations are responsible for progressive cone-rod dystrophy. In addition, progressive cone dystrophy, with or without rod impairment, has also been observed in *CNGB3*-associated ACHM.[60,62,76] In addition, CORD may also be caused by mutations in *GNAT2*, *PDE6C*, or *PDE6H* (Table 1).

2.3 LCA caused by *CNGA3* mutations

A homozygous c.1579C > A (p.L527M) mutation in *CNGA3* was identified in a singleton case with "early-onset nystagmus (at 5 months old), sluggish pupils, no visual response, and nonrecordable ERG at 10 months of age."[21] This mutation was located at a conserved residue in the cGMP-binding domain and was present neither in 200 controls nor in existing databases, such as dbSNP (build 130), 1000 Genomes Project databases, and ExAC. Recently, compound heterozygous mutations, c.955T > C (p.C319R) and c.1573G > A (p.G525S), were identified in another case with LCA.[52] However, clinical data for this case are not provided except for a diagnosis of LCA. In these two cases, the mutations are located at the S6 transmembrane domain and cGMP-binding domain. Further studies are expected to elucidate the association of *CNGA3* mutations with LCA.

2.4 OCT caused by *CNGA3* mutations

Biallelic mutations in *CNGA3* have been identified in at least three cases with OCT, including: (1) G557R and V456fs* in a 16-year-old male with photophobia, nystagmus, visual acuity of 0.25 and 0.2, mildly impaired color vision, and absent cone ERG[13]; (2) A469T and R277C in an 8-year-old female with visual acuity of 0.2 and 0.2, slightly impaired color vision, and absent cone ERG[13]; and (3) Y357C and T565M in a 20-year-old male with mild photophobia, nystagmus, visual acuity of 20/50 and 20/30, normal color vision, normal rod and severely reduced cone ERG, and normal fundus on ophthalmoscopic examination, optical coherence tomography, and fundus autofluorescence.[22] A diagnosis of OCT is hardly applicable for these patients at infancy when the color vision test could not be performed. Apart from *CNGA3*, mutations in several other genes have also been identified in a few patients with OCT or similar phenotype (like bradyopsia), such as *CNGB3*,[77] *GNAT2*,[78] *CEP290*,[79] *RGS9*, and *R9AP*[80,81] (Table 1).

3. Mutation spectrum and genotype–phenotype correlation

3.1 *CNGA3* is one of the most frequently mutated genes

For the six genes associated with ACHM and related retinal diseases, biallelic mutations in *CNGA3* are most frequently reported in Caucasians, accounting for 72.9% families with mutations identified in the six genes (Fig. 2). These data may have bias as one gene might be more frequently analyzed than others. The proportion of mutations in individual genes in Chinese is different from Caucasians although the mutation types and locations are similar between Chinese and Caucasians. One of the most frequently mutated genes in Chinese is *CNGA3* (Fig. 2), which is not only the most common cause of ACHM and CORD in Chinese (Fig. 1) but also one of the most commonly mutated genes in different forms of inherited retinal dystrophies based on whole-exome sequencing on probands from at least 3280 families, targeting exome sequencing on probands from at least 1472 families, and Sanger sequencing on some families. Comparing with the proportion of other genes listed in RetNet (https://sph.uth.edu/retnet/), *CNGA3* is one of the most frequently mutated genes in patients with different forms of retinal dystrophies in our case series.

3.2 Mutation classification and location

To date, at least 144 potential pathogenic mutations in *CNGA3* have been identified in 340 families based on transcript isoform 1 (NM_001298.2) (Table 2), including 104 missense mutations, 18 nonsense mutations, 16 frameshift mutations, 4 inframe mutations, and 2 splicing mutations, accounting for 72.2%, 12.5%, 11.1%, 2.8%, and 1.4%, respectively. All but one mutation located at six of the seven coding exons of *CNGA3* except exon 4 that is absent in transcript isoform 2 (NM_001079878.1), as compared with isoform 1 (NM_001298.2). The remaining one is a splicing mutation located at upstream region of exon 4. In addition, only 9 mutations are located in the first 4 coding exons and their adjacent intronic regions, while all the other 135 mutations are located in the last 3 exons and their adjacent intronic regions, which encode all transmembrane domains and loops, pore region, and the cGMP-binding domains.[82] It is interesting that all the nine mutations at the first four exons are truncated mutations, including four nonsense mutations, four frameshift mutations, and one splicing mutation.

3.3 Pathogenic or benign variants

Most of the mutations reported are predicted to be damaging or causative. Polymorphism affecting coding amino acid is rare in this gene. The evaluation of pathogenicity of these mutations is generally based on informatics analysis as well as the occurrence in phenotype-matched cases with ACHM or CORD. Mutations predicted to result in truncation of the encoded *CNGA3* protein, such as nonsense, frameshift, indels, and splicing site mutations, are very rare in *CNGA3* and most likely to be pathogenic in homozygous or compound heterozygous status, based on previous mutation reports, ExAC database, and our in-house data. Care should be taken for missense variants. A rare missense variant predicted to be benign may not be causative in *CNGA3*. Although biallelic variants in *CNGA3* are very rare in a population with a very limited number of consanguineous marriages, like most Han Chinese, however, a few individuals without *CNGA3*-associated phenotypes may harbor a rare benign variant together with a potential pathogenic variant. Therefore, biallelic variants with an allele of p.T247M or p.E198K may not be causative for ACHM or CORD, especially in Chinese. Additional benign or suspected variants are yet to be identified.

Phenotype-unknown variants listed in databases, such as ExAC (http://exac.broadinstitute.org/), are useful tools in revealing clues for rare benign variants. Segregation analysis in family members as well as database on a large number of phenotype-known normal controls may be of great help in validating benign variant or pathogenic mutations. Functional study on individual variant may provide important evidence of the effect for a missense mutation.[13] Genotype-oriented phenotypic clarification may also provide important information not only for pathogenesis but also for phenotypic expansion.

3.4 Genotype–phenotype correlation

So far, there is no clear evidence to correlate genotype with phenotype. Homozygous missense mutation can cause both ACHM and CORD, while homozygous nonsense or frameshift mutation can cause both ACHM and CORD. Phenotypic difference may not associate with mutation location or associated domain.

CORD caused by *CNGA3* mutation is more frequently seen in Chinese than in Caucasians. This may partly be due to selection of patients involved in analysis of *CNGA3*. The reason for *CNGA3*-associated LCA phenotype is currently unclear, but additional phenotypic clarification of the patients as well as the systemic analysis of other related genes might shed light on the cause of this specific severe phenotype.

3.5 Gene-specific phenotypes

Overall, there are no *CNGA3*-specific ocular or systemic signs or symptoms. However, considering that *CNGA3* is the most commonly mutated gene in ACHM and CORD and even one of the most commonly mutated gene for retinal dystrophy, a Chinese patient with infantile onset photophobia and rapid pendular nystagmus as well as normal ocular structure (especially normal appearing fundus) is highly likely associated with *CNGA3* mutations.

4. Diagnostic guideline

Several symptoms and signs are key for clinical diagnosis for *CNGA3*-associated retinal dystrophy.

4.1 Symptoms

Clinically, an infantile onset rapid horizontal pendular nystagmus with profound photophobia is generally seen in patients with ACHM, CORD, aniridia, and albinism. Iris structure or pigmentary anomaly as well as macular hypoplasia is generally present in aniridia and albinism but is usually absent in ACHM or CORD. Better vision at dark or dim light and complete inability to discriminate colors are the additional signs for *CNGA3*-associated ACHM or CORD.

4.2 Ocular examination

Normal ocular structure (especially normal looking posterior and peripheral fundus) is an important diagnostic sign for *CNGA3*-associated retinopathy. Macular degeneration is very rare in infant or early childhood. Retinal degeneration in mid-peripheral retina usually indicates other forms of retinopathy. For most other forms of CORD, photophobia is less severe or even absent and their visual complaint is usually accompanied with macular changes.

4.3 ERG test

ERG recordings can provide additional diagnostic signs for *CNGA3*-associated retinopathy. Absent or severely reduced cone responses combined with normal or mildly/moderately reduced rod responses in infant are usually seen in *CNGA3*-associated retinopathy or other closely related gene-associated retinopathy in Chinese. Severely reduced rod responses are very rare in *CNGA3*-associated retinopathy so that patients with such changes should be checked further to exclude other genetic factors. We did not see any patient with *CNGA3* mutations who had absent rod responses on ERG recordings.

4.4 Gene test

Biallelic mutations in *CNGA3* are firm evidence for *CNGA3*-associated retinopathy, especially for an infant with a phenotype suggesting ACHM or CORD. For a patient with LCA or other forms of retinopathy, a biallelic variant in *CNGA3* may need additional clarification: (a) the diagnosis may be inappropriate due to lack of key signs or symptoms; (b) additional genetic factors may play roles in the manifestation of the diseases; and (c) although there is no firm evidence to prove this so far, it might be possible that specific mutations in *CNGA3* may cause other phenotypes such as LCA, retinitis pigmentosa, or other forms of retinopathy.

4.5 Biallelic mutations of *CNGA3* in genetic counselling

Clinical genetic counselling is completely different as to the laboratory research. The following tips may be helpful for clinical genetic counselling for a patient or a family with *CNGA3* biallelic mutations.

Biallelic *CNGA3* mutations may not be necessarily pathogenic. It is recommended to check available database or in-house data, especially ethnic-matched controls, in order to exclude ethnic-specific polymorphisms. Special attention is suggested to variants that are predicted to be benign (even very rare) or occur in regions without mutations described before, such as missense variants in the first four exons.

Although it is very rare, biallelic *CNGA3* mutations may not be true if it is detected by next-generation sequencing. Sanger sequencing should be used to confirm the mutations. It is better to identify the origin of a biallelic mutation through segregation analysis of parents or available family members, as in rare circumstance a heterozygous mutation may be incorrectly called as homozygous mutation.

If biallelic mutations identified in a patient with retinopathy other than ACHM or early-onset CORD, it is necessary to recheck the phenotype. If the phenotype is indeed correct, it is possible that there might be other genetic factors that are involved in the pathogenesis other than *CNGA3*. Physician doing clinical genetic counselling may need to inform the patient or their family members that there might be another risk factor that might also cause retinopathy even without *CNGA3* mutations.

The risk for de novo mutations is present almost for everyone. Potential risk of a genetic disease may be reduced through genetic counselling and associated gene test. However, genetic diseases may never be eliminated by gene test and its associated genetic diagnosis. Unknown area always existed so that current information mentioned above may be changed or even corrected with the research progress of the gene and the diseases.

Acknowledgments

We thank all patients and their relatives for their participation in this study. This study was supported by grants from the Key Projects of Guangzhou (201607020013), the Science and Technology Planning Projects of Guangdong (2017B030314025), and the Fundamental Research Funds of the State Key Laboratory of Ophthalmology. Q.Z. is a recipient of National Science Foundation of China for Distinguished Young Scholars.

References

1. Biel M, Zong X, Distler M, et al. Another member of the cyclic nucleotide-gated channel family, expressed in testis, kidney, and heart. *Proc Natl Acad Sci USA*. 1994;91: 3505–3509.
2. Cukkemane A, Seifert R, Kaupp UB. Cooperative and uncooperative cyclic-nucleotide-gated ion channels. *Trends Biochem Sci*. 2011;36:55–64.
3. Sarfare S, Pittler SJ. Focus on molecules: rod photoreceptor cGMP-gated cation channel. *Exp Eye Res*. 2007;85:173–174.
4. Zheng J, Trudeau MC, Zagotta WN. Rod cyclic nucleotide-gated channels have a stoichiometry of three CNGA1 subunits and one CNGB1 subunit. *Neuron*. 2002;36: 891–896.
5. Zhong H, Molday LL, Molday RS, et al. The heteromeric cyclic nucleotide-gated channel adopts a 3A:1B stoichiometry. *Nature*. 2002;420:193–198.
6. Dryja TP, Finn JT, Peng YW, et al. Mutations in the gene encoding the alpha subunit of the rod cGMP-gated channel in autosomal recessive retinitis pigmentosa. *Proc Natl Acad Sci USA*. 1995;92:10177–10181.
7. Bareil C, Hamel CP, Delague V, et al. Segregation of a mutation in CNGB1 encoding the beta-subunit of the rod cGMP-gated channel in a family with autosomal recessive retinitis pigmentosa. *Hum Genet*. 2001;108:328–334.
8. Matveev AV, Quiambao AB, Browning Fitzgerald J, et al. Native cone photoreceptor cyclic nucleotide-gated channel is a heterotetrameric complex comprising both CNGA3 and CNGB3: a study using the cone-dominant retina of Nrl-/- mice. *J Neurochem*. 2008;106:2042–2055.
9. Peng C, Rich ED, Varnum MD. Subunit configuration of heteromeric cone cyclic nucleotide-gated channels. *Neuron*. 2004;42:401–410.
10. Shuart NG, Haitin Y, Camp SS, et al. Molecular mechanism for 3:1 subunit stoichiometry of rod cyclic nucleotide-gated ion channels. *Nat Commun*. 2011;2:457.
11. Kohl S, Marx T, Giddings I, et al. Total colourblindness is caused by mutations in the gene encoding the alpha-subunit of the cone photoreceptor cGMP-gated cation channel. *Nat Genet*. 1998;19:257–259.
12. Sundin OH, Yang JM, Li Y, et al. Genetic basis of total colourblindness among the Pingelapese islanders. *Nat Genet*. 2000;25:289–293.
13. Reuter P, Koeppen K, Ladewig T, et al. Mutations in CNGA3 impair trafficking or function of cone cyclic nucleotide-gated channels, resulting in achromatopsia. *Hum Mutat*. 2008;29:1228–1236.
14. Kohl S, Baumann B, Rosenberg T, et al. Mutations in the cone photoreceptor G-protein alpha-subunit gene GNAT2 in patients with achromatopsia. *Am J Hum Genet*. 2002; 71:422–425.

15. Thiadens AA, den Hollander AI, Roosing S, et al. Homozygosity mapping reveals PDE6C mutations in patients with early-onset cone photoreceptor disorders. *Am J Hum Genet.* 2009;85:240–247.
16. Kohl S, Coppieters F, Meire F, et al. A nonsense mutation in PDE6H causes autosomal-recessive incomplete achromatopsia. *Am J Hum Genet.* 2012;91:527–532.
17. Ansar M, Santos-Cortez RL, Saqib MA, et al. Mutation of ATF6 causes autosomal recessive achromatopsia. *Hum Genet.* 2015;134:941–950.
18. Arbour NC, Zlotogora J, Knowlton RG, et al. Homozygosity mapping of achromatopsia to chromosome 2 using DNA pooling. *Hum Mol Genet.* 1997;6:689–694.
19. Wissinger B, Jagle H, Kohl S, et al. Human rod monochromacy: linkage analysis and mapping of a cone photoreceptor expressed candidate gene on chromosome 2q11. *Genomics.* 1998;51:325–331.
20. Wissinger B, Gamer D, Jagle H, et al. CNGA3 mutations in hereditary cone photoreceptor disorders. *Am J Hum Genet.* 2001;69:722–737.
21. Wang X, Wang H, Cao M, et al. Whole-exome sequencing identifies ALMS1, IQCB1, CNGA3, and MYO7A mutations in patients with Leber congenital amaurosis. *Hum Mutat.* 2011;32:1450–1459.
22. Vincent A, Wright T, Billingsley G, et al. Oligocone trichromacy is part of the spectrum of CNGA3-related cone system disorders. *Ophthalmic Genet.* 2011;32:107–113.
23. Abouelhoda M, Sobahy T, El-Kalioby M, et al. Clinical genomics can facilitate countrywide estimation of autosomal recessive disease burden. *Genet Med.* 2016;18: 1244–1249.
24. Ahuja Y, Kohl S, Traboulsi EI. CNGA3 mutations in two United Arab Emirates families with achromatopsia. *Mol Vis.* 2008;14:1293–1297.
25. Azam M, Collin RW, Malik A, et al. Identification of novel mutations in Pakistani families with autosomal recessive retinitis pigmentosa. *Arch Ophthalmol.* 2011;129: 1377–1378.
26. Azam M, Collin RW, Shah ST, et al. Novel CNGA3 and CNGB3 mutations in two Pakistani families with achromatopsia. *Mol Vis.* 2010;16:774–781.
27. Burgueno-Montanes C, Colunga Cueva M, Costales Alvarez C. A novel mutation in the CNGA3 gene responsible for incomplete achromatopsia. *Arch Soc Esp Oftalmol.* 2014;89:107–109.
28. Carrigan M, Duignan E, Malone CP, et al. Panel-based population next-generation sequencing for inherited retinal degenerations. *Sci Rep.* 2016;6, 33248.
29. Carss KJ, Arno G, Erwood M, et al. Comprehensive rare variant analysis via whole-genome sequencing to determine the molecular pathology of inherited retinal disease. *Am J Hum Genet.* 2017;100:75–90.
30. Chen XT, Huang H, Chen YH, et al. Achromatopsia caused by novel missense mutations in the CNGA3 gene. *Int J Ophthalmol.* 2015;8:910–915.
31. Di Iorio V, Karali M, Brunetti-Pierri R, et al. Clinical and genetic evaluation of a cohort of pediatric patients with severe inherited retinal dystrophies. *Genes (Basel).* 2017;8:280.
32. Doucette L, Green J, Black C, et al. Molecular genetics of achromatopsia in Newfoundland reveal genetic heterogeneity, founder effects and the first cases of Jalili syndrome in North America. *Ophthalmic Genet.* 2013;34:119–129.
33. Eksandh L, Kohl S, Wissinger B. Clinical features of achromatopsia in Swedish patients with defined genotypes. *Ophthalmic Genet.* 2002;23:109–120.
34. Ellingford JM, Barton S, Bhaskar S, et al. Molecular findings from 537 individuals with inherited retinal disease. *J Med Genet.* 2016;53:761–767.
35. Fahim AT, Khan NW, Zahid S, et al. Diagnostic fundus autofluorescence patterns in achromatopsia. *Am J Ophthalmol.* 2013;156:1211–1219.e2.

36. Goto-Omoto S, Hayashi T, Gekka T, et al. Compound heterozygous CNGA3 mutations (R436W, L633P) in a Japanese patient with congenital achromatopsia. *Vis Neurosci.* 2006;23:395–402.
37. Greenberg JP, Sherman J, Zweifel SA, et al. Spectral-domain optical coherence tomography staging and autofluorescence imaging in achromatopsia. *JAMA Ophthalmol.* 2014;132:437–445.
38. Huang L, Xiao X, Li S, et al. Molecular genetics of cone-rod dystrophy in Chinese patients: new data from 61 probands and mutation overview of 163 probands. *Exp Eye Res.* 2016;146:252–258.
39. Johnson S, Michaelides M, Aligianis IA, et al. Achromatopsia caused by novel mutations in both CNGA3 and CNGB3. *J Med Genet.* 2004;41:e20.
40. Kellner U, Wissinger B, Kohl S, et al. Molecular genetic findings in patients with congenital cone dysfunction. Mutations in the CNGA3, CNGB3, or GNAT2 genes. *Ophthalmologe.* 2004;101:830–835.
41. Koeppen K, Reuter P, Kohl S, et al. Functional analysis of human CNGA3 mutations associated with colour blindness suggests impaired surface expression of channel mutants A3(R427C) and A3(R563C). *Eur J Neurosci.* 2008;27:2391–2401.
42. Koeppen K, Reuter P, Ladewig T, et al. Dissecting the pathogenic mechanisms of mutations in the pore region of the human cone photoreceptor cyclic nucleotide-gated channel. *Hum Mutat.* 2010;31:830–839.
43. Kuniyoshi K, Muraki-Oda S, Ueyama H, et al. Novel mutations in the gene for alpha-subunit of retinal cone cyclic nucleotide-gated channels in a Japanese patient with congenital achromatopsia. *Jpn J Ophthalmol.* 2016;60:187–197.
44. Lam K, Guo H, Wilson GA, et al. Identification of variants in CNGA3 as cause for achromatopsia by exome sequencing of a single patient. *Arch Ophthalmol.* 2011;129:1212–1217.
45. Li FF, Huang XF, Chen J, et al. Identification of novel mutations by targeted exome sequencing and the genotype-phenotype assessment of patients with achromatopsia. *J Transl Med.* 2015;13:334.
46. Li L, Chen Y, Jiao X, et al. Homozygosity mapping and genetic analysis of autosomal recessive retinal dystrophies in 144 consanguineous Pakistani families. *Invest Ophthalmol Vis Sci.* 2017;58:2218–2238.
47. Li S, Huang L, Xiao X, et al. Identification of CNGA3 mutations in 46 families: common cause of achromatopsia and cone-rod dystrophies in Chinese patients. *JAMA Ophthalmol.* 2014;132:1076–1083.
48. Liang X, Dong F, Li H, et al. Novel CNGA3 mutations in Chinese patients with achromatopsia. *Br J Ophthalmol.* 2015;99:571–576.
49. Lisowska J, Lisowski L, Kelbsch C, et al. Development of a chromatic pupillography protocol for the first gene therapy trial in patients with CNGA3-linked achromatopsia. *Invest Ophthalmol Vis Sci.* 2017;58:1274–1282.
50. Matet A, Kohl S, Baumann B, et al. Multimodal imaging including semiquantitative short-wavelength and near-infrared autofluorescence in achromatopsia. *Sci Rep.* 2018;8:5665.
51. Nishiguchi KM, Sandberg MA, Gorji N, et al. Cone cGMP-gated channel mutations and clinical findings in patients with achromatopsia, macular degeneration, and other hereditary cone diseases. *Hum Mutat.* 2005;25:248–258.
52. Patel N, Aldahmesh MA, Alkuraya H, et al. Expanding the clinical, allelic, and locus heterogeneity of retinal dystrophies. *Genet Med.* 2016;18:554–562.
53. Riera M, Navarro R, Ruiz-Nogales S, et al. Whole exome sequencing using ion proton system enables reliable genetic diagnosis of inherited retinal dystrophies. *Sci Rep.* 2017;7, 42078.

54. Saqib MA, Awan BM, Sarfraz M, et al. Genetic analysis of four Pakistani families with achromatopsia and a novel S4 motif mutation of CNGA3. *Jpn J Ophthalmol*. 2011;55:676–680.
55. Saqib MA, Nikopoulos K, Ullah E, et al. Homozygosity mapping reveals novel and known mutations in Pakistani families with inherited retinal dystrophies. *Sci Rep*. 2015;5:9965.
56. Schallhorn CS, Granet DB, Ferreyra HA. Electronegative electroretinogram in achromatopsia. *Retin Cases Brief Rep*. 2018;12:143–148.
57. Shaikh RS, Reuter P, Sisk RA, et al. Homozygous missense variant in the human CNGA3 channel causes cone-rod dystrophy. *Eur J Hum Genet*. 2015;23:473–480.
58. Tager J, Kohl S, Birch DG, et al. An early nonsense mutation facilitates the expression of a short isoform of CNGA3 by alternative translation initiation. *Exp Eye Res*. 2018;171:48–53.
59. Taylor RL, Parry NRA, Barton SJ, et al. Panel-based clinical genetic testing in 85 children with inherited retinal disease. *Ophthalmology*. 2017;124:985–991.
60. Thiadens AA, Roosing S, Collin RW, et al. Comprehensive analysis of the achromatopsia genes CNGA3 and CNGB3 in progressive cone dystrophy. *Ophthalmology*. 2010;117:825–830.e1.
61. Thiadens AA, Slingerland NW, Roosing S, et al. Genetic etiology and clinical consequences of complete and incomplete achromatopsia. *Ophthalmology*. 2009;116:1984–1989.e1.
62. Thiadens AA, Somervuo V, van den Born LI, et al. Progressive loss of cones in achromatopsia: an imaging study using spectral-domain optical coherence tomography. *Invest Ophthalmol Vis Sci*. 2010;51:5952–5957.
63. Trankner D, Jagle H, Kohl S, et al. Molecular basis of an inherited form of incomplete achromatopsia. *J Neurosci*. 2004;24:138–147.
64. Varsanyi B, Wissinger B, Kohl S, et al. Clinical and genetic features of Hungarian achromatopsia patients. *Mol Vis*. 2005;11:996–1001.
65. Wiszniewski W, Lewis RA, Lupski JR. Achromatopsia: the CNGB3 p.T383fsX mutation results from a founder effect and is responsible for the visual phenotype in the original report of uniparental disomy 14. *Hum Genet*. 2007;121:433–439.
66. Yang P, Michaels KV, Courtney RJ, et al. Retinal morphology of patients with achromatopsia during early childhood: implications for gene therapy. *JAMA Ophthalmol*. 2014;132:823–831.
67. Zelinger L, Cideciyan AV, Kohl S, et al. Genetics and disease expression in the CNGA3 form of achromatopsia: steps on the path to gene therapy. *Ophthalmology*. 2015;122:997–1007.
68. Zelinger L, Greenberg A, Kohl S, et al. An ancient autosomal haplotype bearing a rare achromatopsia-causing founder mutation is shared among Arab Muslims and Oriental Jews. *Hum Genet*. 2010;128:261–267.
69. Zobor D, Werner A, Stanzial F, et al. The clinical phenotype of CNGA3-related achromatopsia: pretreatment characterization in preparation of a gene replacement therapy trial. *Invest Ophthalmol Vis Sci*. 2017;58:821–832.
70. Thomas MG, McLean RJ, Kohl S, et al. Early signs of longitudinal progressive cone photoreceptor degeneration in achromatopsia. *Br J Ophthalmol*. 2012;96:1232–1236.
71. Kohl S, Varsanyi B, Antunes GA, et al. CNGB3 mutations account for 50% of all cases with autosomal recessive achromatopsia. *Eur J Hum Genet*. 2005;13:302–308.
72. Wang I, Khan NW, Branham K, et al. Establishing baseline rod electroretinogram values in achromatopsia and cone dystrophy. *Doc Ophthalmol*. 2012;125:229–233.
73. Liu C, Varnum MD. Functional consequences of progressive cone dystrophy-associated mutations in the human cone photoreceptor cyclic nucleotide-gated channel CNGA3 subunit. *Am J Physiol Cell Physiol*. 2005;289:C187–C198.

74. Xu J, Morris LM, Michalakis S, et al. CNGA3 deficiency affects cone synaptic terminal structure and function and leads to secondary rod dysfunction and degeneration. *Invest Ophthalmol Vis Sci.* 2012;53:1117–1129.
75. Thapa A, Morris L, Xu J, et al. Endoplasmic reticulum stress-associated cone photoreceptor degeneration in cyclic nucleotide-gated channel deficiency. *J Biol Chem.* 2012;287:18018–18029.
76. Khan NW, Wissinger B, Kohl S, et al. CNGB3 achromatopsia with progressive loss of residual cone function and impaired rod-mediated function. *Invest Ophthalmol Vis Sci.* 2007;48:3864–3871.
77. Andersen MK, Christoffersen NL, Sander B, et al. Oligocone trichromacy: clinical and molecular genetic investigations. *Invest Ophthalmol Vis Sci.* 2010;51:89–95.
78. Rosenberg T, Baumann B, Kohl S, et al. Variant phenotypes of incomplete achromatopsia in two cousins with GNAT2 gene mutations. *Invest Ophthalmol Vis Sci.* 2004;45:4256–4262.
79. Roosing S, Cremers FPM, Riemslag FCC, et al. A rare form of retinal dystrophy caused by hypomorphic nonsense mutations in CEP290. *Genes (Basel).* 2017;8.
80. Michaelides M, Li Z, Rana NA, et al. Novel mutations and electrophysiologic findings in RGS9- and R9AP-associated retinal dysfunction (Bradyopsia). *Ophthalmology.* 2010;117:120–127.e1.
81. Strauss RW, Dubis AM, Cooper RF, et al. Retinal architecture in RGS9- and R9AP-associated retinal dysfunction (Bradyopsia). *Am J Ophthalmol.* 2015;160:1269–1275-e1.
82. Zagotta WN, Siegelbaum SA. Structure and function of cyclic nucleotide-gated channels. *Annu Rev Neurosci.* 1996;19:235–263.

CHAPTER TWO

Arrestin mutations: Some cause diseases, others promise cure

Vsevolod V. Gurevich[1], Eugenia V. Gurevich
Department of Pharmacology, Vanderbilt University, Nashville, TN, United States
[1]Corresponding author: e-mail address: vsevolod.gurevich@vanderbilt.edu

Contents

Abstract

Arrestins play a key role in homologous desensitization of G protein-coupled receptors (GPCRs) and regulate several other vital signaling pathways in cells. Considering the critical roles of these proteins in cellular signaling, surprisingly few disease-causing mutations in human arrestins were described. Most of these are loss-of-function mutations of visual arrestin-1 that cause excessive rhodopsin signaling and hence night blindness. Only one dominant arrestin-1 mutation was discovered so far. It reduces the thermal stability of the protein, which likely results in photoreceptor death via unfolded protein response. In case of the two nonvisual arrestins, only polymorphisms were described, some of which appear to be associated with neurological disorders and altered response to certain treatments. Structure–function studies revealed several ways of enhancing arrestins' ability to quench GPCR signaling. These enhanced arrestins have potential as tools for gene therapy of disorders associated with excessive signaling of mutant GPCRs.

Abbreviations

ERK extracellular signal-regulated kinase
GPCR G protein-coupled receptor
GRP78/BIP glucose-regulated protein of 78 kDa/immunoglobulin binding protein
JNK c-Jun N-terminal kinase
MAP kinase mitogen-activated protein kinase
WT wild type
β2AR β2-adrenergic receptor

Progress in Molecular Biology and Translational Science, Volume 161
ISSN 1877-1173
https://doi.org/10.1016/bs.pmbts.2018.09.004

1. Arrestins in mammals: Few subtypes, many functions

The first arrestin described was the visual subtype (systematic name arrestin-1[a]). In fact, it was discovered not for its biological role as we know it today, but as an antigen causing uveitis,[1] hence it was called S-antigen. *Arrestin-1* gene is called *SAG* (abbreviation for S-antigen) to this day. A year later Kuhn described a "48-kDa" protein in the retina that, along with the visual G protein transducin and rhodopsin kinase, binds rhodopsin in a light-dependent manner.[2] A few years later Kuhn found that rhodopsin phosphorylation (also discovered by his group many years earlier[3]) greatly facilitates the binding of the 48-kDa protein to rhodopsin.[4] The fact that S-antigen and 48-kDa protein are one and the same was established only in mid-1980s.[5,6] Subsequent studies by Dr. Kuhn's group showed that the binding of the 48-kDa protein to phosphorylated rhodopsin quenches light-dependent activation of photoreceptor phosphodiesterase,[7] which was the first evidence for the role of this protein in "arresting" rhodopsin signaling, i.e., in receptor desensitization. Subsequent discoveries of the kinase that selectively phosphorylates agonist-activated β2-adrenergic receptor (β2AR)[8] and the first nonvisual arrestin[9] demonstrated that two-step desensitization, phosphorylation of activated GPCR by a specific receptor kinase followed by arrestin binding to the active phosphorylated receptor, is a common feature of the signaling systems driven by GPCRs (reviewed in ref. 10). Mammals express between ~500 (dolphins) and ~3400 (elephants) different GPCRs (http://sevens.cbrc.jp/), but only 4 arrestin subtypes.[11,12] Two of these four are specialized visual: arrestin-1, expressed at very high levels in both rod and cone photoreceptors,[13–16] and cone-specific arrestin-4, constituting ~2% of total arrestin complement in cones.[13] The other two nonvisual subtypes, arrestin-2 and -3, are ubiquitously expressed and apparently interact with hundreds of different GPCRs.[9,17–19] In most cell types arrestin-2 greatly outnumbers arrestin-3,[20,21] which explains why arrestin-2[9] was cloned years before arrestin-3.[22–24]

Arrestin-1 was shown to suppress G protein activation by phosphorylated light-activated rhodopsin via direct competition.[25,26] Nonvisual arrestins were found to perform exactly the same function, the only

[a] We use systematic names of arrestin proteins, where the number after the dash indicates the order of cloning: arrestin-1 (historic names S-antigen, 48 kDa protein, visual or rod arrestin), arrestin-2 (β-arrestin or β-arrestin1), arrestin-3 (β-arrestin2 or hTHY-ARRX), and arrestin-4 (cone or X-arrestin).

difference being that they prefer other GPCRs over rhodopsin.[22,27] Soon after their discovery, both nonvisual arrestins were shown to bind clathrin[28] and clathrin adaptor AP-2,[29] two key players in receptor internalization via coated pits. Many other nonreceptor binding partners of the two nonvisual arrestins were subsequently discovered. Interestingly, some apparently bind both arrestin-2 and -3, whereas others selectively interact with only one of the nonvisual subtypes.[30] Arrestins were implicated in the activation of c-Src[31] and all three major MAP kinase classes, JNK,[32] ERK,[33] and p38,[34] shown to bind cAMP phosphodiesterase,[35] calmodulin,[36] microtubules,[37,38] ubiquitin ligases Mdm2,[39] AIP4,[40] parkin,[41] and many other signaling proteins (reviewed in refs. 11,42,43). Thus, it appears that arrestins are at the crossroads of numerous vital signaling pathways in the cell.

2. Naturally occurring mutations in visual arrestins

Critical role of arrestin proteins in the visual system was confirmed by abnormally prolonged photoresponses in mice lacking arrestin-1,[44] rhodopsin kinase,[45] or expressing rhodopsin without phosphorylation sites necessary for its binding[46] or with insufficient number of these sites.[47] Indeed, in humans frameshift mutation in *SAG* was found to underlie Oguchi disease, a form of stationary night blindness.[48,49] This mutation is a deletion of an adenine in codon 309, which causes a frameshift and premature termination 10 residues later.[48,50] Thus, the resulting protein lacks about half of the C-domain[51] and is unlikely to fold or become functional. Interestingly, the same mutation in the *SAG* gene caused autosomal recessive retinitis pigmentosa (a form of retinal degeneration) in some cases.[50] Although it is not known why the phenotype is so different, certain mechanisms are conceivable. All eukaryotic cells have special mechanisms of "garbage disposal," at mRNA and protein levels. One of these is nonsense-mediated mRNA decay, i.e., the elimination of mutant or incorrectly spliced mRNA containing premature stop codons.[52–54] The other is the degradation of misfolded and unfolded proteins via ubiquitin-proteasome system (reviewed in ref. 55). Insufficient activity of either can exacerbate the problem caused by this frameshift in the *SAG* gene. In particular, incomplete elimination of faulty mRNA would make the cell produce too much unfolded mutant arrestin-1, overwhelming ubiquitin-proteasome system and causing cell death via unfolded protein response (see below). This is particularly likely considering that arrestin-1 is the second most abundant protein in rods after rhodopsin,[14–16] so that even if a small fraction of defective mRNA escapes the nucleus and is translated in the cytoplasm, the output of unfolded protein would be large.

Loss-of-function mutations are recessive, i.e., the production of normal arrestin-1 encoded by the second undamaged allele is usually sufficient for health of rod photoreceptors. This is consistent with the results obtained in mice, where hemizygous knockout animals and even the animals expressing arrestin-1 at ~5% of wild-type (WT) level had perfectly healthy photoreceptors and normal rate of photoresponse shutoff after moderately bright flashes.[16,56]

Recently a dominant mutation in *SAG*, C147F, was discovered[57] (Fig. 1A). Its dominant nature suggested that the mutant itself, rather than the absence of functional arrestin-1, causes the damage that WT arrestin-1 protein generated by the second allele cannot prevent. Although cysteine in this position is conserved in all arrestin subtypes[11,12] and in highly homologous bovine and mouse arrestin-1 and it was shown to be important for arrestin function,[58] in and of itself this did not explain dominant nature of the C147F mutation. Careful inspection of the structure of arrestin-1 in its basal conformation,[51] as well as in complex with rhodopsin[59,60] suggested that the replacements of a cysteine with a relatively small side chain by a bulky phenylalanine in this position can affect protein folding (Fig. 1B).[57] This idea was tested experimentally, and the results turned out to be more complex. It was shown that C147F arrestin-1 folds, but demonstrates much lower thermal stability than the WT protein, denaturing within hours even at physiological temperature

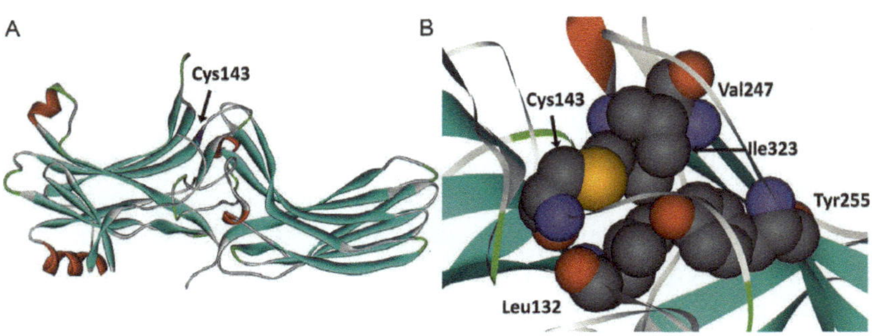

Fig. 1 Dominant mutation in visual arrestin-1 causing retinal degeneration. (A) The position of Cys-147 in the arrestin-1 molecule (panels A and B are based on the crystal structure of bovine arrestin-1, PDB ID:1CF1,[51] so the homologous bovine Cys-143 is shown). (B) Cys-143 and its neighbors (indicated) are shown as CPK models, with the atoms colored, as follows: carbon, *gray*; oxygen, *red*; nitrogen, *blue*; sulfur, *yellow*. Note close packing, which would make the introduction of much bulkier side chain destabilizing.

of 37°C.[61] Similar loss of thermal stability was demonstrated in case of C147I and C147L, but not with C147A or C147V mutants with smaller side chains, clearly indicating that the bulk of the side chain underlies the problem.[61] The expression of this mutant in photoreceptor-derived 661W cells,[62] in contrast to WT human arrestin-1, was shown to induce unfolded protein response.[61] Thus, considering extremely high expression of arrestin-1 in rods,[14–16] it appears that denaturing C147F mutant overwhelms proteasome system and induces photoreceptor death via unfolded protein response (Fig. 2). It is quite likely that a mutation in the stop codon of *SAG*, which added an extra of 25 amino acids, causes late-onset hereditary retinal degeneration in dogs[64] via a similar mechanism.

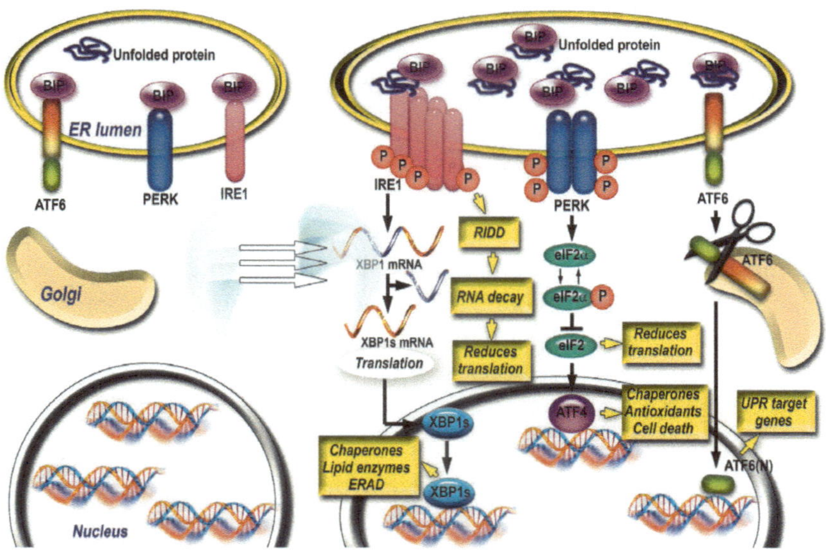

Fig. 2 Mutant arrestin-1 induces unfolded protein response. Experimental data suggest that C147F mutant is significantly less stable than WT arrestin-1.[61] As arrestin-1 is the second most abundant protein in the rod photoreceptors after rhodopsin, its unfolding likely induces unfolded protein response, which eventually leads to rod death.[63] An increase in GRP78/BIP, which manifests the first step of the unfolded protein response, was detected in photoreceptor-derived 661W cells expressing this mutant, but not in cells expressing WT human arrestin-1.[61] The schematic is based on the mechanisms described earlier (see ref. 63 and references therein). Abbreviations: *ATF4*, activating transcription factor 4; *ATF6*, activating transcription factor 6; *BIP*, same as GPR78, a major ER chaperone, a.k.a. 78-kDa glucose-regulated protein; *eIF2α*, eukaryotic translation initiation factor 2α; *ER*, endoplasmic reticulum; *ERAD*, ER-associated protein degradation; *IRE1*, inositol-requiring enzyme 1; *PERK*, PKR-like ER kinase; *RIDD*, regulated IRE1-dependent decay of mRNA; *UPR*, unfolded protein response; *XBP1*, X-box binding protein 1.

Interestingly, no disease-causing mutations or polymorphisms were so far reported in cone-specific *arrestin-4* (*ARR4*), even though it appears to be necessary for proper function of cone photoreceptors.[13]

3. Nonvisual arrestins: Unexpectedly few associations with unclear functional significance

Numerous residues are highly conserved in arrestin evolution,[11,12] which suggests their functional importance. Yet not a single disease-related change in the amino acid sequence has so far been described in human nonvisual arrestins. Several single nucleotide polymorphisms in the arrestin-2 and -3 genes (*ARRB1* and *ARRB2*), which include synonymous mutations in the coding sequence and single nucleotide polymorphisms (SNPs) in the noncoding elements of their mRNAs and in the introns, have been described. The associations of these polymorphisms with various conditions have been extensively tested.

Polymorphism in the promoter region of the *ARRB1* gene was found to be associated with the treatment outcome with antidepressant mirtazapine in Korean patients with major depressive disorder.[65] One haplotype, which includes SNPs in the promoter region and the first intron, was found to be associated with the remission status following several weeks of mirtazapine treatment. Two SNPs in *ARRB2*, one in an intron and one in 3′-untranslated region (3′-UTR), were reported to be associated with the effects of antidepressant monotherapy in a group of mostly Caucasian patients with major depression.[66] Interestingly, the same SNP in the 3′-UTR of *ARRB2* mRNA, rs4790694, was found to be associated with two measures of nicotine dependence in European Americans.[67] There was also one major haplotype in both *ARRB2* and *ARRB1* that showed positive association with the indices of nicotine dependence. Interestingly, no such associations were seen in African Americans.[67] A study of association between polymorphism in the *ARRB2* gene and alcohol dependence found no such association in Caucasians.[68] Four SNPs spanning the entire gene were examined (rs4790694 was not included). These human data are somewhat at odds with the finding in animals that demonstrated the existence of an *Arrb2* haplotype consisting of six SNPs in different parts of the gene and one insertion (in the promoter region) that was specifically associated with ethanol preference in rats.[69] The variant haplotype in the ethanol-preferring rat line confers higher arrestin-3 expression both at mRNA and protein levels in several brain regions, but no change in the arrestin-3 protein sequence.

An extensive role of arrestins in signaling as well as in regulation of the GPCR responsiveness drove the investigation of genetic association of arrestin polymorphisms and the effects of addictive drugs. However, investigation of seven *ARRB2* SNPs, including rs4790694, failed to reveal significant association of any individual SNP or haplotype with cocaine or opioid dependence in European Americans.[70] However, polymorphism in *ARRB2* was found to impact the outcome of methadone substitution therapy in Caucasian patients addicted to opioids. The study examined four SNPs in the *ARRB2* gene, from the promoter to 3′-UTR, three of which, in intron 1, exon 11 (synonymous Ser280Ser), and 3′-UTR, were found to be significantly associated with response to methadone therapy, although no association was seen with opioid addiction.[71] These three SNPs formed a haplotype block, and patients homozygous for variant alleles in the block carried an almost threefold higher risk of being nonresponders to the methadone therapy.

A study of four *ARRB2* SNPs failed to detect an association between any of them and schizophrenia in Japanese patients.[72] However, three out of five SNPs (synonymous rs1045280 in exon 11 and rs2036657 and rs4790694 in 3′-UTR) were significantly associated with methamphetamine use disorders such as methamphetamine-induced psychosis. Additionally, a significant association was found between rs1045280 Ser280Ser SNP and tardive dyskinesia, a motor complication induced by long-term treatment with typical antipsychotics.[73]

Overall, these association studies revealed a lot less than one would expect. Considering vital role of nonvisual arrestins in many cellular functions, one would expect to find numerous disease-associated mutations and polymorphisms. However, very few have been described so far. The major problem with interpretation of the existing findings is, of course, the lack of obvious functional significance of these genetic variants. None of the SNPs change the proteins sequence of arrestins. All known SNPs in the exons are synonymous. Many SNPs found associated with human diseases are located in introns or in 3′-UTR, with unknown functional impact. Out of all *ARRB2* SNPs examined so far, only one SNP, rs34230287 (−159C/T), located in the gene promoter, has been shown to affect the promoter activity and *ARRB2* expression: the C variant confers a significantly higher promoter activity and is associated with higher level of *ARRB2* mRNA in Caucasians.[74] However, this SNP was either not examined or, when it was, showed no association with the outcome measures.[71]

Although in most cases this was not tested, the most likely effect of SNPs in coding and noncoding regions is a change in expression. The most extreme change in protein expression is produced by the knockout of its gene. If we look at mouse in vivo studies, the phenotypes detected in single nonvisual arrestin knockouts are fairly mild: mice lacking either *Arrb1* or *Arrb2* are overall normal, albeit demonstrate altered functional responses such as increased sensitivity to adrenergic stimulation of the heart[75] or enhanced locomotor responsiveness to amphetamine[76] in *Arrb1* knockout mice or enhanced morphine analgesia[77] and reduced locomotor response to amphetamine[76,78] and morphine[79] in *Arrb2* knockout mice. Interestingly, simultaneous knockout of both nonvisual subtypes is embryonic lethal because of lung and heart development problems.[80,81] The elimination of the only nonvisual arrestin, *kurtz*, in *Drosophila* is also embryonic lethal,[82] even though in flies kurtz inhibits proproliferative MAP kinase ERK,[83] in contrast to mammalian cells, where nonvisual arrestins facilitate ERK activation via c-Src[31] and by scaffolding of c-Raf1-MEK1-ERK1/2 cascade.[33,84]

One reason for these observations in knockout mice and humans with SNPs is that the two nonvisual subtypes in mammals can partially compensate for one another. Mild phenotypes of *Arrb1* and *Arrb2* knockout mice appear to support this argument. However, in certain aspects these two subtypes are quite different and engage distinct sets of signaling proteins.[30] A recent study suggests that arrestin-2 and arrestin-3 play distinct roles in amphetamine-induced hyperlocomotion and that their effects are dose dependent.[76] While the mechanistic basis for the distinct roles of the two nonvisual arrestins was not determined in that study, there are several well-known functional differences: arrestin-3 has higher affinity than arrestin-2 for clathrin[28] and several GPCRs,[18,80] arrestin-3 is less selective than arrestin-2 for the active phosphorylated form of cognate GPCRs,[85] and arrestin-3, but not arrestin-2, facilitates JNK3 activation in cells.[32,86,87] The latter difference was preserved even in short arrestin-derived N-terminal peptides.[88] One of these differences, or some functional difference that has not been elucidated yet, might underlie distinct roles of arrestin-2 and arrestin-3 in amphetamine-induced hyperlocomotion.[76] However, mild phenotypes of single subtype knockout mice suggest that either nonvisual arrestin can perform most of the biologically important functions of the other subtype.

4. Enhanced arrestins: Compensation of excessive GPCR activity

Numerous mutations in various GPCRs were reported to cause different kinds of pathological conditions in humans (reviewed in refs. 89,90).

Some of these mutations are loss of function, so conceptually the strategy for gene therapy is clear: the delivery of a normal coding sequence of a functional GPCR to the affected cells should solve the problem. However, other identified disease-causing mutations are gain of function, where the mutant gene encodes an overactive receptor. Unlike recessive loss-of-function, gain-of-function mutations are dominant, as perfectly normal second allele cannot suppress excessive signaling by the mutant. One possible therapeutic strategy is the expression of an arrestin with enhanced ability to dampen the signaling, so that, on balance, the signal might become near normal. This compensational approach was so far tested only in rod photoreceptors, where the important GPCR is rhodopsin, which is quenched exclusively by arrestin-1.[86] Extensive structure–function studies of arrestin-1 (reviewed in refs. 85,91) yielded several enhanced versions that bind both phosphorylated and unphosphorylated light-activated rhodopsin tighter than parental WT protein (Fig. 3). As usual, translation of the in vitro findings to the in vivo situation yielded both good and bad news. On the positive side, transgenic expression of enhanced arrestin-1 in mouse rods defective in rhodopsin phosphorylation improves the morphology of photoreceptors, their functional performance, and results in signal shutoff that is much faster than with WT arrestin-1.[86] Importantly, facilitation of the shutoff is also documented in mice with defects of rhodopsin phosphorylation expressing normal complement or WT arrestin-1, similar to human patients with rhodopsin mutations.[92] However, neither parameter in "compensated" rods came even close to that of normal rods where rhodopsin was phosphorylated and quenched by WT arrestin-1.[86] As far as the visual system is concerned, with

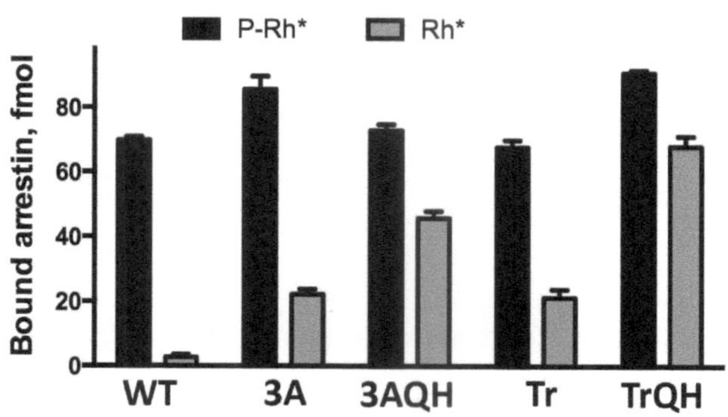

Fig. 3 Phosphorylation dependence of the binding of WT arrestin-1 and its enhanced mutants. The binding in the in vitro assay of translated radiolabeled arrestins to purified phosphorylated (P-Rh*) and unphosphorylated (Rh*) light-activated rhodopsin (performed as described[17]) is shown.

its unrivaled single photon sensitivity and subsecond shutoff of the response, these results suggested that while in principle the strategy is working, more powerful enhanced mutants are necessary for better compensation.[93] However, something that works only partially in rods might be sufficient to fully compensate for excessive signaling in any other GPCR-driven system, where the sensitivity is much lower and the shutoff takes minutes, rather than 200–300 ms.

The mechanism of GPCR binding is well conserved in the arrestin family,[91] so both nonvisual arrestins can be enhanced by the mutations homologous to those that preactivate arrestin-1.[94–96] Yet in case of nonvisual GPCRs there is another catch: both nonvisual arrestins are quite promiscuous and bind pretty much every GPCR tested.[17–19,97] As virtually every cell in the body expresses numerous GPCR subtypes, only one of which is an overactive mutant in patients, introduction of a promiscuous enhanced nonvisual arrestin would likely suppress the signaling not only of the "bad guy" but of all the other perfectly normal receptors coexpressed in the same cell. This is likely to cause unwanted side effects. Thus, to make the same compensational strategy usable, mutant nonvisual arrestins with narrow receptor specificity are needed. While the receptor–arrestin interface is extensive, involving many residues on the concave sides of both arrestin domains,[59,98–100] only select few appear to play a role in receptor preference[100] (Fig. 4). The results of the targeted manipulation of these putative receptor-discriminator residues suggest that the construction of nonvisual

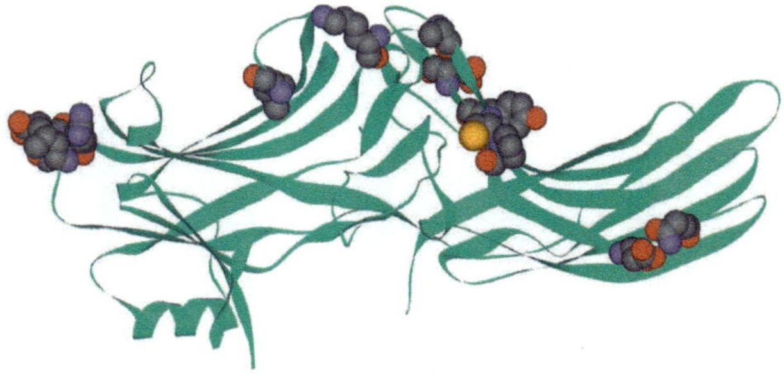

Fig. 4 Arrestin-3 residues that determine receptor preference. The residues found to change receptor preference of nonvisual arrestin-3 in the in-cell BRET-based assay[19,101–103] are shown as CPK models on the crystal structure of the basal state of bovine arrestin-3 (PDB ID: 3P2D[85]), with the atoms colored, as follows: *gray*, carbon; *red*, oxygen; *blue*, nitrogen; *yellow*, sulfur. Note the localization of these residues on the concave sides of both arrestin domains, with clustering on the central crest of the receptor-binding side of the molecule.

arrestins with narrow receptor specificity is feasible: certain double mutants demonstrate 50–60-fold preference for some GPCRs over others,[19] and a number of mutations differentially affect arrestin interactions with distinct functional forms of several GPCRs.[101–103] Generally speaking, preactivating mutations often reduce receptor specificity of arrestins.[94,96] However, this is so in case of active phosphorylated receptors, whereas the same mutations enhance binding only to cognate unphosphorylated GPCRs.[94,96] Moreover, enhanced mutants actually compete with GRKs, suppressing the phosphorylation of GPCRs they target.[95] Thus, while receptor specificity of the mutants combining preactivating mutations with those that make them specific for particular GPCRs needs to be tested experimentally, this compensational approach appears to be a feasible method to rein in excessive signaling by many disease-causing gain-of-function receptor mutants.

5. Conclusions

Several mutations in *SAG* were reported to cause visual disorders in humans and dogs. Loss-of-function mutations are usually recessive, so that only compound heterozygotes are affected, whereas mutations causing misfolding and/or protein instability are dominant and cause severe disorders, such as retinal degeneration. Unexpectedly, no mutations in either nonvisual arrestin were associated with any disease, while polymorphisms in noncoding regions and synonymous base substitutions in exons were found to be associated with the response to treatment of several neurological disorders. The phenotypes of single nonvisual arrestin knockout in mice are mild, whereas simultaneous knockout of both, as well as the knockout of the only nonvisual arrestin in *Drosophila*, causes embryonic lethality. Thus, the most logical explanation of mild effects of single knockouts in mammals is that arrestin-2 and arrestin-3 can compensate for the missing subtype, taking over each other's duties. Preactivated arrestin mutants with enhanced ability to quench GPCR signaling appear to be a viable tool for the gene therapy of disorders caused by gain-of-function mutations in GPCRs. The use of this strategy to suppress excessive signaling by nonvisual GPCRs requires engineering of receptor-specific nonvisual arrestin mutants.

Acknowledgments

This work was supported in part by NIH RO1 grants EY011500, GM077561, and GM109955 (the latter two RO1s were replaced by R35 GM122491) (V.V.G.), and NS065868 and DA030103 (E.V.G.).

References

1. Wacker WB, Donoso LA, Kalsow CM, Yankeelov Jr JA, Organisciak DT. Experimental allergic uveitis. Isolation, characterization, and localization of a soluble uveitopathogenic antigen from bovine retina. *J Immunol*. 1977;119:1949–1958.
2. Kuhn H. Light-regulated binding of rhodopsin kinase and other proteins to cattle photoreceptor membranes. *Biochemistry*. 1978;17:4389–4395.
3. Kuhn H, Dreyer WJ. Light dependent phosphorylation of rhodopsin by ATP. *FEBS Lett*. 1972;20:1–6.
4. Kuhn H, Hall SW, Wilden U. Light-induced binding of 48-kDa protein to photoreceptor membranes is highly enhanced by phosphorylation of rhodopsin. *FEBS Lett*. 1984;176:473–478.
5. Pfister C, Chabre M, Plouet J, et al. Retinal S antigen identified as the 48K protein regulating light-dependent phosphodiesterase in rods. *Science*. 1985;228:891–893.
6. Pfister C, Dorey C, Vadot E, et al. Identification of the so-called 48 K protein that interacts with illuminated rhodopsin in retinal rods, and the retinal S antigen, inductor of experimental autoimmune uveoretinitis. *C R Acad Sci III*. 1984;299:261–265.
7. Wilden U, Hall SW, Kühn H. Phosphodiesterase activation by photoexcited rhodopsin is quenched when rhodopsin is phosphorylated and binds the intrinsic 48-kDa protein of rod outer segments. *Proc Natl Acad Sci USA*. 1986;83:1174–1178.
8. Benovic JL, DeBlasi A, Stone WC, Caron MG, Lefkowitz RJ. Beta-adrenergic receptor kinase: primary structure delineates a multigene family. *Science*. 1989; 246:235–240.
9. Lohse MJ, Benovic JL, Codina J, Caron MG, Lefkowitz RJ. Beta-arrestin: a protein that regulates beta-adrenergic receptor function. *Science*. 1990;248:1547–1550.
10. Carman CV, Benovic JL. G-protein-coupled receptors: turn-ons and turn-offs. *Curr Opin Neurobiol*. 1998;8:335–344.
11. Gurevich EV, Gurevich VV. Arrestins are ubiquitous regulators of cellular signaling pathways. *Genome Biol*. 2006;7:236.
12. Indrischek H, Prohaska SJ, Gurevich VV, Gurevich EV, Stadler PF. Uncovering missing pieces: duplication and deletion history of arrestins in deuterostomes. *BMC Evol Biol*. 2017;17:163.
13. Nikonov SS, Brown BM, Davis JA, et al. Mouse cones require an arrestin for normal inactivation of phototransduction. *Neuron*. 2008;59:462–474.
14. Strissel KJ, Sokolov M, Trieu LH, Arshavsky VY. Arrestin translocation is induced at a critical threshold of visual signaling and is superstoichiometric to bleached rhodopsin. *J Neurosci*. 2006;26:1146–1153.
15. Hanson SM, Gurevich EV, Vishnivetskiy SA, Ahmed MR, Song X, Gurevich VV. Each rhodopsin molecule binds its own arrestin. *Proc Natl Acad Sci USA*. 2007;104: 3125–3128.
16. Song X, Vishnivetskiy SA, Seo J, Chen J, Gurevich EV, Gurevich VV. Arrestin-1 expression in rods: balancing functional performance and photoreceptor health. *Neuroscience*. 2011;174:37–49.
17. Gurevich VV, Dion SB, Onorato JJ, et al. Arrestin interaction with G protein-coupled receptors. Direct binding studies of wild type and mutant arrestins with rhodopsin, b2-adrenergic, and m2 muscarinic cholinergic receptors. *J Biol Chem*. 1995;270: 720–731.
18. Barak LS, Ferguson SS, Zhang J, Caron MG. A beta-arrestin/green fluorescent protein biosensor for detecting G protein-coupled receptor activation. *J Biol Chem*. 1997;272:27497–27500.
19. Gimenez LE, Vishnivetskiy SA, Baameur F, Gurevich VV. Manipulation of very few receptor discriminator residues greatly enhances receptor specificity of non-visual arrestins. *J Biol Chem*. 2012;287:29495–29505.

20. Gurevich EV, Benovic JL, Gurevich VV. Arrestin2 and arrestin3 are differentially expressed in the rat brain during postnatal development. *Neuroscience.* 2002;109: 421–436.
21. Gurevich EV, Benovic JL, Gurevich VV. Arrestin2 expression selectively increases during neural differentiation. *J Neurochem.* 2004;91:1404–1416.
22. Attramadal H, Arriza JL, Aoki C, et al. Beta-arrestin2, a novel member of the arrestin/ beta-arrestin gene family. *J Biol Chem.* 1992;267:17882–17890.
23. Sterne-Marr R, Gurevich VV, Goldsmith P, et al. Polypeptide variants of beta-arrestin and arrestin3. *J Biol Chem.* 1993;268:15640–15648.
24. Rapoport B, Kaufman KD, Chazenbalk GD. Cloning of a member of the arrestin family from a human thyroid cDNA library. *Mol Cell Endocrinol.* 1992;84:R39–R43.
25. Wilden U. Duration and amplitude of the light-induced cGMP hydrolysis in vertebrate photoreceptors are regulated by multiple phosphorylation of rhodopsin and by arrestin binding. *Biochemistry.* 1995;34:1446–1454.
26. Krupnick JG, Gurevich VV, Benovic JL. Mechanism of quenching of phototransduction. Binding competition between arrestin and transducin for phosphorhodopsin. *J Biol Chem.* 1997;272:18125–18131.
27. Lohse MJ, Andexinger S, Pitcher J, et al. Receptor-specific desensitization with purified proteins. Kinase dependence and receptor specificity of beta-arrestin and arrestin in the beta 2-adrenergic receptor and rhodopsin systems. *J Biol Chem.* 1992;267:8558–8564.
28. Goodman Jr OB, Krupnick JG, Santini F, et al. Beta-arrestin acts as a clathrin adaptor in endocytosis of the beta2-adrenergic receptor. *Nature.* 1996;383:447–450.
29. Laporte SA, Oakley RH, Zhang J, et al. The β2-adrenergic receptor/arrestin complex recruits the clathrin adaptor AP-2 during endocytosis. *Proc Natl Acad Sci USA.* 1999;96:3712–3717.
30. Xiao K, McClatchy DB, Shukla AK, et al. Functional specialization of beta-arrestin interactions revealed by proteomic analysis. *Proc Natl Acad Sci USA.* 2007;104:12011–12016.
31. Luttrell LM, Ferguson SS, Daaka Y, et al. Beta-arrestin-dependent formation of beta2 adrenergic receptor-Src protein kinase complexes. *Science.* 1999;283:655–661.
32. McDonald PH, Chow CW, Miller WE, et al. Beta-arrestin 2: a receptor-regulated MAPK scaffold for the activation of JNK3. *Science.* 2000;290:1574–1577.
33. Luttrell LM, Roudabush FL, Choy EW, et al. Activation and targeting of extracellular signal-regulated kinases by beta-arrestin scaffolds. *Proc Natl Acad Sci USA.* 2001;98:2449–2454.
34. Bruchas MR, Macey TA, Lowe JD, Chavkin C. Kappa opioid receptor activation of p38 MAPK is GRK3- and arrestin-dependent in neurons and astrocytes. *J Biol Chem.* 2006;281:18081–18089.
35. Perry SJ, Baillie GS, Kohout TA, et al. Targeting of cyclic AMP degradation to beta 2-adrenergic receptors by beta-arrestins. *Science.* 2002;298:834–836.
36. Wu N, Hanson SM, Francis DJ, et al. Arrestin binding to calmodulin: a direct interaction between two ubiquitous signaling proteins. *J Mol Biol.* 2006;364:955–963.
37. Nair KS, Hanson SM, Kennedy MJ, Hurley JB, Gurevich VV, Slepak VZ. Direct binding of visual arrestin to microtubules determines the differential subcellular localization of its splice variants in rod photoreceptors. *J Biol Chem.* 2004;279: 41240–41248.
38. Hanson SM, Cleghorn WM, Francis DJ, et al. Arrestin mobilizes signaling proteins to the cytoskeleton and redirects their activity. *J Mol Biol.* 2007;368:375–387.
39. Shenoy SK, McDonald PH, Kohout TA, Lefkowitz RJ. Regulation of receptor fate by ubiquitination of activated beta 2-adrenergic receptor and beta-arrestin. *Science.* 2001;294:1307–1313.

40. Bhandari D, Trejo J, Benovic JL, Marchese A. Arrestin-2 interacts with the ubiquitin-protein isopeptide ligase atrophin-interacting protein 4 and mediates endosomal sorting of the chemokine receptor CXCR4. *J Biol Chem*. 2007;282:36971–36979.
41. Ahmed MR, Zhan X, Song X, Kook S, Gurevich VV, Gurevich EV. Ubiquitin ligase parkin promotes Mdm2-arrestin interaction but inhibits arrestin ubiquitination. *Biochemistry*. 2011;50:3749–3763.
42. Gurevich VV, Gurevich EV. The structural basis of arrestin-mediated regulation of G protein-coupled receptors. *Pharmacol Ther*. 2006;110:465–502.
43. Peterson YK, Luttrell LM. The diverse roles of arrestin scaffolds in G protein-coupled receptor signaling. *Pharmacol Rev*. 2017;69:256–297.
44. Xu J, Dodd RL, Makino CL, Simon MI, Baylor DA, Chen J. Prolonged photoresponses in transgenic mouse rods lacking arrestin. *Nature*. 1997;389:505–509.
45. Chen CK, Burns ME, Spencer M, et al. Abnormal photoresponses and light-induced apoptosis in rods lacking rhodopsin kinase. *Proc Natl Acad Sci USA*. 1999;96: 3718–3722.
46. Chen J, Makino CL, Peachey NS, Baylor DA, Simon MI. Mechanisms of rhodopsin inactivation in vivo as revealed by a COOH-terminal truncation mutant. *Science*. 1995;267:374–377.
47. Mendez A, Burns ME, Roca A, et al. Rapid and reproducible deactivation of rhodopsin requires multiple phosphorylation sites. *Neuron*. 2000;28:153–164.
48. Fuchs S, Nakazawa M, Maw M, Tamai M, Oguchi Y, Gal A. A homozygous 1-base pair deletion in the arrestin gene is a frequent cause of Oguchi disease in Japanese. *Nat Genet*. 1995;10:360–362.
49. Nakazawa M, Wada Y, Fuchs S, Gal A, Tamai M. Oguchi disease: phenotypic characteristics of patients with the frequent 1147delA mutation in the arrestin gene. *Retina*. 1997;17:17–22.
50. Nakazawa M, Wada Y, Tamai M. Arrestin gene mutations in autosomal recessive retinitis pigmentosa. *Arch Ophthalmol*. 1998;16:498–501.
51. Hirsch JA, Schubert C, Gurevich VV, Sigler PB. The 2.8 A crystal structure of visual arrestin: a model for arrestin's regulation. *Cell*. 1999;97:257–269.
52. Frischmeyer PA, Dietz HC. Nonsense-mediated mRNA decay in health and disease. *Hum Mol Genet*. 1999;8:1893–1900.
53. Hilleren P, Parker R. Mechanisms of mRNA surveillance in eukaryotes. *Annu Rev Genet*. 1999;33:229–260.
54. Hentze MW, Kulozik AE. A perfect message: RNA surveillance and nonsense-mediated decay. *Cell*. 1999;96:307–310.
55. Berner N, Reutter KR, Wolf DH. Protein quality control of the endoplasmic reticulum and ubiquitin-proteasome-triggered degradation of aberrant proteins: yeast pioneers the path. *Annu Rev Biochem*. 2018;87:751–782.
56. Cleghorn WM, Tsakem EL, Song X, et al. Progressive reduction of its expression in rods reveals two pools of arrestin-1 in the outer segment with different roles in photoresponse recovery. *PLoS One*. 2011;6, e22797.
57. Sullivan LS, Bowne SJ, Koboldt DC, et al. A novel dominant mutation in SAG, the arrestin-1 gene, is a common cause of retinitis pigmentosa in Hispanic families in the Southwestern United States. *Invest Ophthalmol Vis Sci*. 2017;58:2774–2784.
58. Vishnivetskiy SA, Lee RJ, Zhou XE, et al. Functional role of the three conserved cysteines in the N domain of visual arrestin-1. *J Biol Chem*. 2017;292:12496–12502.
59. Kang Y, Zhou XE, Gao X, et al. Crystal structure of rhodopsin bound to arrestin determined by femtosecond X-ray laser. *Nature*. 2015;523:561–567.
60. Zhou XE, He Y, de Waal PW, et al. Structural identification of phosphorylation codes for arrestin recruitment by G protein-coupled receptors. *Cell*. 2017;170:457–469.

61. Vishnivetskiy SA, Sullivan LS, Bowne SJ, Daiger SP, Gurevich EV, Gurevich VV. Molecular defects of the disease-causing human arrestin-1 C147F mutant. *Invest Ophthalmol Vis Sci*. 2018;59:13–20.
62. Al-Ubaidi MR, Font RL, Quiambao AB, et al. Bilateral retinal and brain tumors in transgenic mice expressing simian virus 40 large T antigen under control of the human interphotoreceptor retinoid-binding protein promoter. *J Cell Biol*. 1992;119:1681–1687.
63. Kim R, Emi M, Tanabe K, Murakami S. Role of the unfolded protein response in cell death. *Apoptosis*. 2006;11:5–13.
64. Goldstein O, Jordan JA, Aguirre GD, Acland GM. A non-stop S-antigen gene mutation is associated with late onset hereditary retinal degeneration in dogs. *Mol Vis*. 2013;19:1871–1884.
65. Chang HS, Won ES, Lee HY, Ham BJ, Kim YG, Lee MS. Association of ARRB1 polymorphisms with the risk of major depressive disorder and with treatment response to mirtazapine. *J Psychopharmacol*. 2015;29:615–622.
66. Petit AC, El Asmar K, David DJ, et al. The association of β-arrestin2 polymorphisms with response to antidepressant treatment in depressed patients. *Prog Neuro-psychopharmacol Biol Psychiatry*. 2018;81:74–79.
67. Sun D, Ma JZ, Payne TJ, Li MD. Beta-arrestins 1 and 2 are associated with nicotine dependence in European American smokers. *Mol Psychiatry*. 2008;13:398–406.
68. Oneda B, Preisig M, Dobrinas M, Eap CB. Lack of association between genetic polymorphisms of ARRB2 and alcohol dependence in a Caucasian population. *Alcohol Alcohol*. 2010;45:590–591.
69. Björk K, Rimondini R, Hansson AC, et al. Modulation of voluntary ethanol consumption by beta-arrestin 2. *FASEB J*. 2008;22:2552–2560.
70. Ambrose-Lanci LM, Vaswani M, Clarke TK, et al. Association study of the β-arrestin 2 gene (ARRB2) with opioid and cocaine dependence in a European-American population. *Psychiatr Genet*. 2012;22:141–145.
71. Oneda B, Crettol S, Bochud M, et al. β-Arrestin2 influences the response to methadone in opioid-dependent patients. *Pharm J*. 2010;11:258.
72. Ikeda M, Ozaki N, Suzuki T, et al. Possible association of beta-arrestin 2 gene with methamphetamine use disorder, but not schizophrenia. *Genes Brain Behav*. 2007;6: 107–112.
73. Liou YJ, Wang YC, Chen JY, et al. The coding-synonymous polymorphism rs1045280 (Ser280Ser) in beta-arrestin 2 (ARRB2) gene is associated with tardive dyskinesia in Chinese patients with schizophrenia. *Eur J Neurol*. 2008;15:1406–1408.
74. Zhang X, He J-Q, Ding L, Paré PD, Sandford AJ. Promoter polymorphism and expression of β-arrestin 2 in neutrophils. *Clin Chim Acta*. 2007;385:79–80.
75. Conner DA, Mathier MA, Mortensen RM, et al. β-Arrestin1 knockout mice appear normal but demonstrate altered cardiac responses to β-adrenergic stimulation. *Circ Res*. 1997;81:1021–1026.
76. Zurkovsky L, Sedaghat K, Ahmed MR, Gurevich VV, Gurevich EV. Arrestin-2 and arrestin-3 differentially modulate locomotor responses and sensitization to amphetamine. *Neuropharmacology*. 2017;121:20–29.
77. Bohn LM, Lefkowitz RJ, Gainetdinov RR, Peppel K, Caron MG, Lin FT. Enhanced morphine analgesia in mice lacking beta-arrestin 2. *Science*. 1999;286:2495–2498.
78. Beaulieu JM, Sotnikova TD, Marion S, Lefkowitz RJ, Gainetdinov RR, Caron MG. An Akt/beta-arrestin 2/PP2A signaling complex mediates dopaminergic neurotransmission and behavior. *Cell*. 2005;122:261–273.
79. Bohn LM, Gainetdinov RR, Sotnikova TD, et al. Enhanced rewarding properties of morphine, but not cocaine, in beta(arrestin)-2 knock-out mice. *J Neurosci*. 2003;23: 10265–10273.

80. Kohout TA, Lin FS, Perry SJ, Conner DA, Lefkowitz RJ. Beta-arrestin 1 and 2 differentially regulate heptahelical receptor signaling and trafficking. *Proc Natl Acad Sci USA*. 2001;98:1601–1606.
81. Zhang M, Liu X, Zhang Y, Zhao J. Loss of betaarrestin1 and betaarrestin2 contributes to pulmonary hypoplasia and neonatal lethality in mice. *Dev Biol*. 2010;339:407–417.
82. Roman G, He J, Davis RL. kurtz, a novel nonvisual arrestin, is an essential neural gene in *Drosophila*. *Genetics*. 2000;155:1281–1295.
83. Tipping M, Kim Y, Kyriakakis P, Tong M, Shvartsman SY, Veraksa A. β-Arrestin Kurtz inhibits MAPK and Toll signalling in *Drosophila* development. *EMBO J*. 2010;29:3222–3235.
84. Coffa S, Breitman M, Spiller BW, Gurevich VV. A single mutation in arrestin-2 prevents ERK1/2 activation by reducing c-Raf1 binding. *Biochemistry*. 2011;50: 6951–6958.
85. Zhan X, Gimenez LE, Gurevich VV, Spiller BW. Crystal structure of arrestin-3 reveals the basis of the difference in receptor binding between two non-visual arrestins. *J Mol Biol*. 2011;406:467–478.
86. Song X, Coffa S, Fu H, Gurevich VV. How does arrestin assemble MAPKs into a signaling complex? *J Biol Chem*. 2009;284:685–695.
87. Seo J, Tsakem EL, Breitman M, Gurevich VV. Identification of arrestin-3-specific residues necessary for JNK3 kinase activation. *J Biol Chem*. 2011;286:27894–27901.
88. Zhan X, Stoy H, Kaoud TS, et al. Peptide mini-scaffold facilitates JNK3 activation in cells. *Sci Rep*. 2016;6:21025.
89. Schoneberg T, Schulz A, Biebermann H, Hermsdorf T, Rompler H, Sangkuhl K. Mutant G-protein-coupled receptors as a cause of human diseases. *Pharmacol Ther*. 2004;104:173–206.
90. Stoy H, Gurevich VV. How genetic errors in GPCRs affect their function: possible therapeutic strategies. *Genes Dis*. 2015;2:108–132.
91. Gurevich VV, Gurevich EV. The molecular acrobatics of arrestin activation. *Trends Pharmacol Sci*. 2004;25:105–111.
92. Samaranayake S, Song X, Vishnivetskiy SA, Chen J, Gurevich EV, Gurevich VV. Enhanced mutant compensates for defects in rhodopsin phosphorylation in the presence of endogenous arrestin-1. *Front Mol Neurosci*. 2018;11:203.
93. Vishnivetskiy SA, Chen Q, Palazzo MC, et al. Engineering visual arrestin-1 with special functional characteristics. *J Biol Chem*. 2013;288:11741–11750.
94. Kovoor A, Celver J, Abdryashitov RI, Chavkin C, Gurevich VV. Targeted construction of phosphorylation-independent β-arrestin mutants with constitutive activity in cells. *J Biol Chem*. 1999;274:6831–6834.
95. Pan L, Gurevich EV, Gurevich VV. The nature of the arrestin x receptor complex determines the ultimate fate of the internalized receptor. *J Biol Chem*. 2003;278: 11623–11632.
96. Celver J, Vishnivetskiy SA, Chavkin C, Gurevich VV. Conservation of the phosphate-sensitive elements in the arrestin family of proteins. *J Biol Chem*. 2002;277:9043–9048.
97. Oakley RH, Laporte SA, Holt JA, Caron MG, Barak LS. Differential affinities of visual arrestin, βarrestin1, and βarrestin2 for G protein-coupled receptors delineate two major classes of receptors. *J Biol Chem*. 2000;275:17201–17210.
98. Hanson SM, Francis DJ, Vishnivetskiy SA, et al. Differential interaction of spin-labeled arrestin with inactive and active phosphorhodopsin. *Proc Natl Acad Sci USA*. 2006;103:4900–4905.
99. Hanson SM, Gurevich VV. The differential engagement of arrestin surface charges by the various functional forms of the receptor. *J Biol Chem*. 2006;281:3458–3462.

100. Vishnivetskiy SA, Gimenez LE, Francis DJ, et al. Few residues within an extensive binding interface drive receptor interaction and determine the specificity of arrestin proteins. *J Biol Chem*. 2011;286:24288–24299.
101. Chen Q, Perry NA, Vishnivetskiy SA, et al. Structural basis of arrestin-3 activation and signaling. *Nat Commun*. 2017;8:1427.
102. Gimenez LE, Babilon S, Wanka L, Beck-Sickinger AG, Gurevich VV. Mutations in arrestin-3 differentially affect binding to neuropeptide Y receptor subtypes. *Cell Signal*. 2014;26:1523–1531.
103. Prokop S, Perry NA, Vishnivetskiy SA, et al. Differential manipulation of arrestin-3 binding to basal and agonist-activated G protein-coupled receptors. *Cell Signal*. 2017;36:98–107.

CHAPTER THREE

Mutations in *GPR101* as a potential cause of X-linked acrogigantism and acromegaly

Zhi-Shuai Hou[a], Ya-Xiong Tao[a,b,*]
[a]Department of Anatomy, Physiology and Pharmacology, College of Veterinary Medicine, Auburn University, Auburn, AL, United States
[b]Center for Neuroscience Initiative, Auburn University, Auburn, AL, United States
*Corresponding author: e-mail address: taoyaxi@auburn.edu

Contents

Abstract

X-linked acrogigantism (XLAG) is a recently described early-onset gigantism due to *GPR101* duplication that induces growth hormone (GH) oversecretion. GPR101, which belongs to Family A rhodopsin-like family of G protein-coupled receptors, is predominantly expressed in hypothalamus and pituitary, suggesting that GPR101 might be important in regulating diverse functions such as energy balance and reproduction. Most mammalian GPR101s have extremely long third intracellular loops (ICL3); however, zebrafish GPR101 has a much shorter ICL3, but a longer C-terminus. GnRH-(1–5), a GnRH metabolite, can modulate the hypothalamus–pituitary–gonad axis and cancer cell

Progress in Molecular Biology and Translational Science, Volume 161
ISSN 1877-1173
https://doi.org/10.1016/bs.pmbts.2018.10.003

migration via activating GPR101. GPR101 couples to both $G\alpha_s$ and $G\alpha_i$ proteins. *GPR101* duplication has a causative role in XLAG, while *GPR101* variants, especially c.924G > C (E308D), located at ICL3, are attributed to acromegaly. Some *GPR101* mutations that are associated with a small proportion of pituitary tumors without GH oversecretion have also been identified recently. This chapter will summarize studies on GPR101, including its molecular cloning and tissue distribution, physiology, pharmacology, and pathophysiology.

Abbreviations

3′/5′-UTR	3′/5′-untranslated region
α-, β-, and γ-MSH	α-, β-, and γ-melanocyte stimulating hormone
AIP	aryl hydrocarbon receptor-interacting protein
Chinese hamster ovary	CHO cells
CRE-luciferase reporter	cAMP response element-luciferase reporter
E_2	17β-estradiol
EGF	epidermal growth factor
EGFR	epidermal growth factor receptor
ERE	estrogen-response element
ExAC database	Exome Aggregation Consortium database
GH	growth hormone
GHRH	growth hormone-releasing hormone
GnRH	gonadotropin-releasing hormone
GPCR	G protein-coupled receptor
HEK293	human embryonic kidney 293 cells
HPG	hypothalamus–pituitary–gonad
ICL	intracellular loop
IGF-1	insulin-like growth factor-1
MAS	McCune-Albright syndrome
MC3R	melanocortin-3 receptor
MC4R	melanocortin-4 receptor
MEN	multiple endocrine neoplasia
MMP	matrix metalloprotease
mPOA	medial preoptic area
NPY	neuropeptide-Y
OVX	ovariectomy
POMC	proopiomelanocortin
SST	somatostatin
XLAG	X-linked acrogigantism

1. Introduction

Somatic growth is orchestrated by a multihormone crosstalk involving growth hormone-releasing hormone (GHRH), somatostatin (SST), growth hormone (GH), and insulin-like growth factor-1 (IGF-1).[1] GHRH and SST are secreted from the hypothalamus and regulate GH secretion from the pituitary. After secretion into the bloodstream, GH triggers the liver, muscle, and other organs to produce IGF-1, a hormone that plays an important role in stimulating the growth of bones, muscles, and other tissues. However, oversecretion of GH results in excessive amounts of IGF-1, causing abnormal growth of muscles and bones, enlargement of organs, as well as other symptoms and characteristics of gigantism and acromegaly. When GH oversecretion begins in childhood, prior to epiphyseal fusion, the patient usually has gigantism, a rare pathological condition leading to many changes in the body including excessive body growth, enlarged organ and muscle, and delayed puberty. Gigantism in most patients is caused by either tumors or lesions of pituitary, causing GH dyshomeostasis or further lesions in the hypothalamus leading to GHRH hypersecretion.[2] Acromegaly occurs if GH hypersecretion begins after epiphyseal fusion. Most patients with acromegaly display enlarged hands and feet, coarsened and enlarged facial features, and coarse, oily, and thickened skin resulting from GH oversecretion.

Around 50% of gigantism cases are caused by an unknown genetic etiology, and loss-of-function mutations in *AIP* (aryl hydrocarbon receptor-interacting protein) (29%–41%) and *GPR101* (G protein-coupled receptor 101) duplications (8%–10%) can explain most of the remaining 50% cases. Multiple endocrine neoplasia (MEN) syndromes also contribute a small portion of gigantism. Patients with *AIP* mutations or MEN syndromes usually begin abnormal growth during late childhood or early adolescence. However, patients with X-linked acrogigantism (XLAG) usually display an early-onset pituitary gigantism starting in the first 6 months of life.[2,3] XLAG is a genomic disorder due to microduplications of the Xq26.3 chromosomal region encompassing four genes (*CD40LG*, *ARHGEF6*, *RBMX*, and *GPR101*).[4,5] A recent study indicates the causative role of *GPR101* by identifying one XLAG patient who harbors a microduplication encompassing *GPR101* only, but *CD40LG*, *ARHGEF6*, and *RBMX* are not duplicated.[6]

2. Molecular cloning and localization of GPR101

2.1 Molecular cloning

GPR101 belongs to Family A rhodopsin-like family of G protein-coupled receptors (GPCRs).[7] Human *GPR101* was initially identified on the X chromosome by Lee et al. in 2001 as an orphan GPCR with mRNA transcripts that are predominantly expressed in the caudate putamen and hypothalamus.[8] Human GPR101 and another orphan receptor GPR161 share the most homology. Moreover, human *GPR101* shares about 30% identities in transmembrane domain with adrenergic, serotonin, muscarinic, and dopamine receptors.[8] Murine *Gpr101* gene also localizes in chromosome X, a region that is syntenic with the location of human *GPR101*.

Bioinformatical prediction shows that human *GPR101* transcripts consist of three regions, including a 215 bp-long region 1 for the putative 5′-untranslated region (UTR); a 2380 bp-long region 2 that includes the coding sequence as well as a part of the putative 3′-UTR; and a 1901 bp-long region 3 that represents most of the putative 3′-UTR.[9] In vitro studies identified four human *GPR101* isoforms with different 5′-UTRs, resulting from alternative splicing and a common 3′-UTR that extends for 6.1 kbp. Isoform 1 has a single exon while the other three isoforms have two exons.[9] Isoform 1 is the predominant isoform in human pituitary, whereas isoforms 2, 3, and 4 are minor transcripts. Isoform 1 expression is significantly higher in pituitary tumors due to *GPR101* duplication; however, its expression is relatively low in normal pituitary tissue and pituitary tumors without *GPR101* duplication.[4,9] Isoforms 2, 3, and 4 are only expressed in *GPR101*-duplicated pituitary tumors at low levels.[9]

2.2 Orthologue alignment

GPR101 has been identified or predicted in multiple mammalian species. Sequence alignment and phylogenetic analysis of GPR101 in different species are shown in Fig. 1A–C. Human *GPR101* is predicted to encode a protein of 508 amino acids, while rat and mouse *Gpr101* are predicted to encode proteins of 508 and 511 amino acids, respectively.[2,10] Zebrafish *gpr101* only encodes a protein of 437 amino acids. In most mammalian species, GPR101 contains an extremely long intracellular loop (ICL) 3 (Fig. 2A); however, zebrafish GPR101 has a much shorter ICL3 but a longer C-terminus[2,9] (Fig. 2B).

2.3 Tissue distribution

2.3.1 *GPR101* expression in human

In human, different *GPR101* expression profiles exist in adult and childhood. *GPR101* expression is very low in almost all adult human tissues except for specific brain regions.[9] Trivellin et al. analyzed *GPR101* mRNA levels in a panel of 24 different adult human brain regions, demonstrating that the highest level of *GPR101* expression is observed in the nucleus accumbens of the hypothalamus and relatively high *GPR101* expression

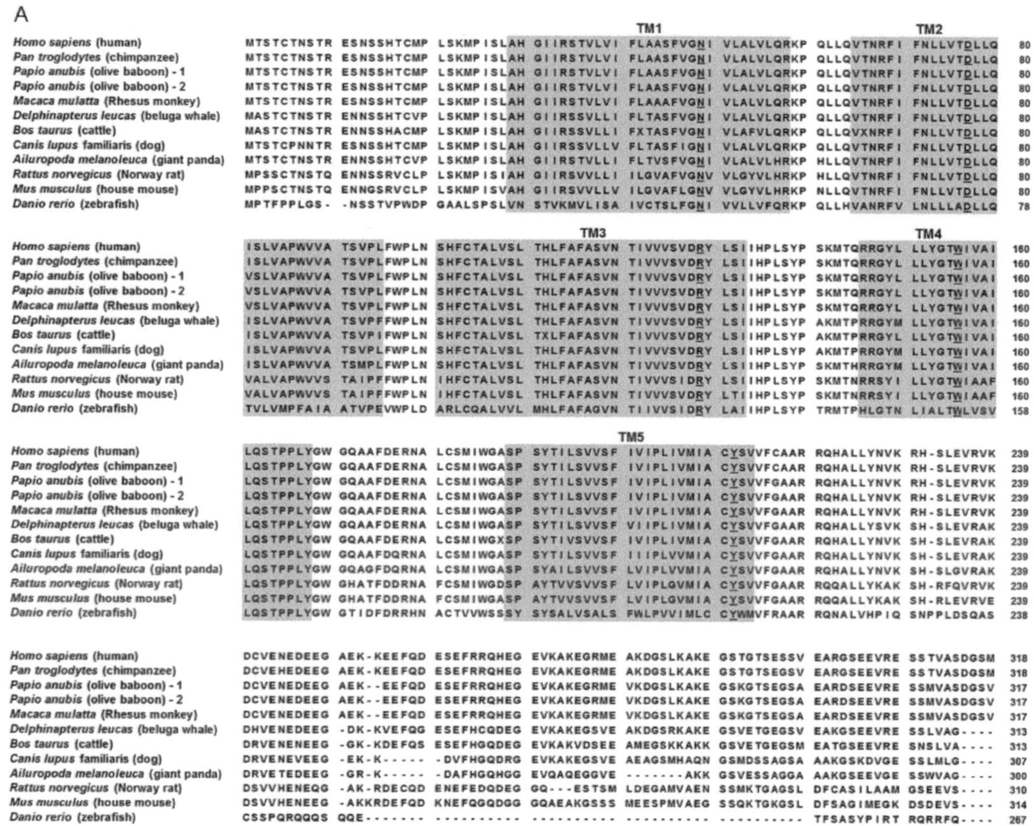

Fig. 1 Comparison of GPR101 amino acid sequences between GPR101s from human and other species. (A) Alignment of the amino acid sequences between human GPR101 and GPR101s cloned from other species. The transmembrane domains are shaded in *gray*. (B) Homology and (C) phylogenetic analysis of GPR101 amino acid sequences from different species. NCBI reference numbers for the proteins are *Homo sapiens* (human), NP_473362.1; *Mus musculus* (mouse), NP_001028532.1; *Rattus norvegicus* (Rat), NP_001101728.1; *Pan troglodytes* (chimpanzee), XP_016799818.1; *Papio anubis* (olive baboon)-1, XP_021788424.1; *Papio anubis* (olive baboon)-2, XP_021788423.1; *Bos taurus* (cattle), XP_024845112.1; *Delphinapterus leucas* (beluga whale), XP_022410026.1; *Macaca mulatta* (Rhesus monkey), XP_014983799.1; *Canis lupus* (dog), XP_022271864.1; *Ailuropoda melanoleuca* (giant panda), XP_002919967.1; *Danio rerio* (zebrafish), XP_005173281.1.

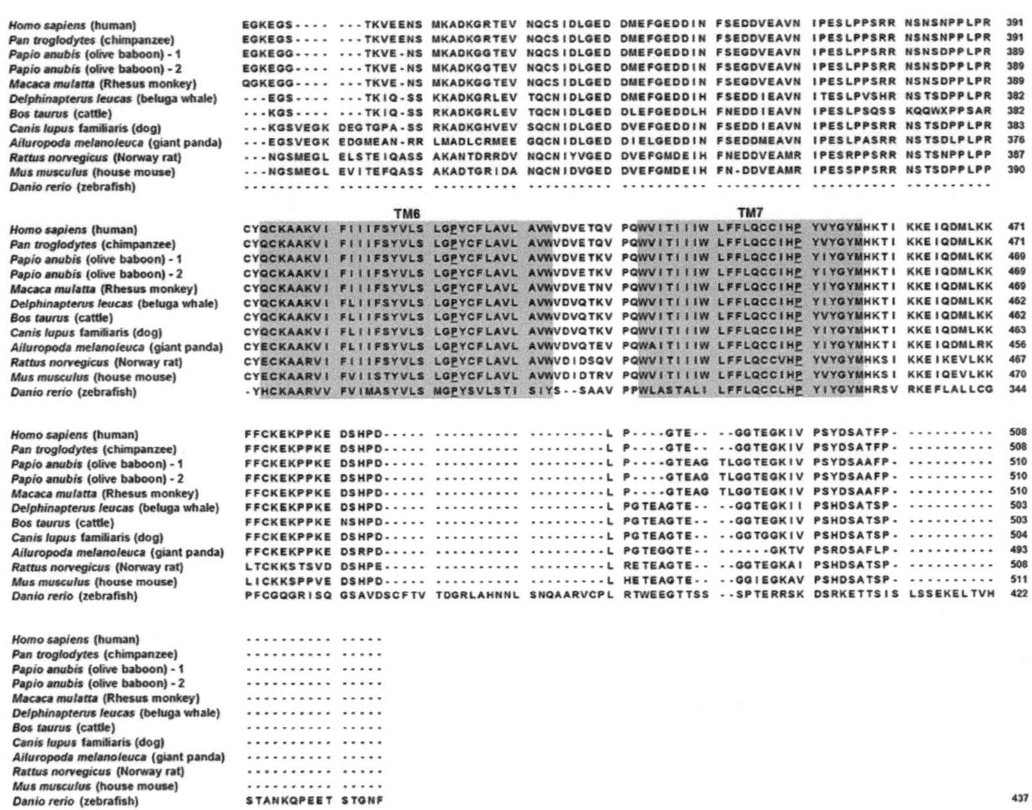
Homo sapiens (human) EGKEGS---- ---TKVEENS MKADKGRTEV NQCSIDLGED DMEFGEDDIN FSEDDVEAVN IPESLPPSRR NSNSNPPLPR 391
Pan troglodytes (chimpanzee) EGKEGS---- ---TKVEENS MKADKGRTEV NQCSIDLGED DMEFGEDDIN FSEDDVEAVN IPESLPPSRR NSNSNPPLPR 391
Papio anubis (olive baboon) - 1 EGKEGG---- ---TKVE-NS MKADKGGTEV NQCSIDLGED DMEFGEDDIN FSEGDVEAVN IPESLPPSRR NSNSDPPLPR 389
Papio anubis (olive baboon) - 2 EGKEGG---- ---TKVE-NS MKADKGGTEV NQCSIDLGED DMEFGEDDIN FSEGDVEAVN IPESLPPSRR NSNSDPPLPR 389
Macaca mulatta (Rhesus monkey) QGKEGG---- ---TKVE-NS MKADKGGTEV NQCSIDLGED DMEFGEDDIN FSEGDVEAVN IPESLPPSRR NSNSDPPLPR 389
Delphinapterus leucas (beluga whale) ---EGS---- ---TKIQ-SS KKADKGRLEV TQCNIDLGED DMEFGEDDIH FSEDDIEAVN ITESLPVSHR NSTSDPPLPR 382
Bos taurus (cattle) ---KGS---- ---TKIQ-SS RKADKGRLEV TQCNIDLGED DLEFGEDDLH FNEDDIEAVN IPESLPSQSS KQQWXPPSAR 382
Canis lupus familiaris (dog) ---KGSVEGK DEGTGPA-SS RKADKGHVEV SQCNIDLGED DVEFGEDDIN FSEDDIEAVN IPESLPPSRR NSTSDPPLPR 383
Ailuropoda melanoleuca (giant panda) ---EGSVEGK EDGMEAN-RR LMADLCRMEE GQCNIDLGED DIELGEDDIN FSEDDMEAVN IPESLPASRR NSTSDLPLPR 376
Rattus norvegicus (Norway rat) ---NGSMEGL ELSTEIQASS AKANTDRRDV NQCNIYVGED DVEFGMDEIH FNEDDVEAMR IPESRPPSRR NSTSNPPLPP 387
Mus musculus (house mouse) ---NGSMEGL EVITEFQASS AKADTGRIDA NQCNIDVGED DVEFGMDEIH FN-DDVEAMR IPESSPPSRR NSTSDPPLPP 390
Danio rerio (zebrafish) ---------- ---------- ---------- ---------- ---------- ---------- ---------- ----------
TM6
TM7
Homo sapiens (human) CYQCKAAKVI FIIIFSYVLS LGPYCFLAVL AVWVDVETQV PQWVITIIIW LFFLQCCIHP YVYGYMHKTI KKEIQDMLKK 471
Pan troglodytes (chimpanzee) CYQCKAAKVI FIIIFSYVLS LGPYCFLAVL AVWVDVETQV PQWVITIIIW LFFLQCCIHP YVYGYMHKTI KKEIQDMLKK 471
Papio anubis (olive baboon) - 1 CYQCKAAKVI FIIIFSYVLS LGPYCFLAVL AVWVDVETKV PQWVITIIIW LFFLQCCIHP YIYGYMHKTI KKEIQDMLKK 469
Papio anubis (olive baboon) - 2 CYQCKAAKVI FIIIFSYVLS LGPYCFLAVL AVWVDVETKV PQWVITIIIW LFFLQCCIHP YIYGYMHKTI KKEIQDMLKK 469
Macaca mulatta (Rhesus monkey) CYQCKAAKVI FIIIFSYVLS LGPYCFLAVL AVWVDVETNV PQWVITIIIW LFFLQCCIHP YIYGYMHKTI KKEIQDMLKK 469
Delphinapterus leucas (beluga whale) CYQCKAAKVI FLIIFSYVLS LGPYCFLAVL AVWVDVQTKV PQWVITIIIW LFFLQCCIHP YIYGYMHKTI KKEIQDMLKK 462
Bos taurus (cattle) CYQCKAAKVI FLIIFSYVLS LGPYCFLAVL AVWVDVQTKV PQWVITIIIW LFFLQCCIHP YVYGYMHKTI KKEIQDMLKK 462
Canis lupus familiaris (dog) CYQCKAAKVI FLIIFSYVLS LGPYCFLAVL AVWVDVQTKV PQWVITIIIW LFFLQCCIHP YIYGYMHKTI KKEIQDMLKK 463
Ailuropoda melanoleuca (giant panda) CYECKAAKVI FLIIFSYVLS LGPYCFLAVL AVWVDVQTEV PQWAITIIIW LFFLQCCIHP YIYGYMHKTI KKEIQDMLRK 456
Rattus norvegicus (Norway rat) CYECKAARVI FIIIFSYVLS LGPYCFLAVL AVWVDIDSQV PQWVITIIIW LFFLQCCVHP YVYGYMHKSI KKEIKEVLKK 467
Mus musculus (house mouse) CYECKAARVI FVIISTYVLS LGPYCFLAVL AVWVDIDTRV PQWVITIIIW LFFLQCCIHP YVYGYMHKSI KKEIQEVLKK 470
Danio rerio (zebrafish) -YHCKAARVV FVIMASYVLS MGPYSVLSTI SIYS--SAAV PPWLASTALI LFFLQCCLHP YIYGYMHRSV RKEFLALLCG 344
Homo sapiens (human) FFCKEKPPKE DSHPD----- ---------- ---------L P----GTE-- --GGTEGKIV PSYDSATFP- ---------- 508
Pan troglodytes (chimpanzee) FFCKEKPPKE DSHPD----- ---------- ---------L P----GTE-- --GGTEGKIV PSYDSATFP- ---------- 508
Papio anubis (olive baboon) - 1 FFCKEKPPKE DSHPD----- ---------- ---------L P----GTEAG TLGGTEGKIV PSYDSAAFP- ---------- 510
Papio anubis (olive baboon) - 2 FFCKEKPPKE DSHPD----- ---------- ---------L P----GTEAG TLGGTEGKIV PSYDSAAFP- ---------- 510
Macaca mulatta (Rhesus monkey) FFCKEKPPKE DSHPD----- ---------- ---------L P----GTEAG TLGGTEGKIV PSYDSAAFP- ---------- 510
Delphinapterus leucas (beluga whale) FFCKEKPPKE DSHPD----- ---------- ---------L PGTEAGTE-- --GGTEGKII PSHDSATSP- ---------- 503
Bos taurus (cattle) FFCKEKPPKE NSHPD----- ---------- ---------L PGTEAGTE-- --GGTEGKIV PSHDSATSP- ---------- 503
Canis lupus familiaris (dog) FFCKEKPPKE DSHPD----- ---------- ---------L PGTEAGTE-- --GGTGGKIV PSHDSATSP- ---------- 504
Ailuropoda melanoleuca (giant panda) FFCKEKPPKE DSRPD----- ---------- ---------L PGTEGGTE-- ------GKTV PSRDSAFLP- ---------- 493
Rattus norvegicus (Norway rat) LTCKKSTSVD DSHPE----- ---------- ---------L RETEAGTE-- --GGTEGKAI PSHDSATSP- ---------- 508
Mus musculus (house mouse) LICKKSPPVE DSHPD----- ---------- ---------L HETEAGTE-- --GGIEGKAV PSHDSATSP- ---------- 511
Danio rerio (zebrafish) PFCGQGRISQ GSAVDSCFTV TDGRLAHNNL SNQAARVCPL RTWEEGTTSS --SPTERRSK DSRKETTSIS LSSEKELTVH 422
Homo sapiens (human) ---------- -----
Pan troglodytes (chimpanzee) ---------- -----
Papio anubis (olive baboon) - 1 ---------- -----
Papio anubis (olive baboon) - 2 ---------- -----
Macaca mulatta (Rhesus monkey) ---------- -----
Delphinapterus leucas (beluga whale) ---------- -----
Bos taurus (cattle) ---------- -----
Canis lupus familiaris (dog) ---------- -----
Ailuropoda melanoleuca (giant panda) ---------- -----
Rattus norvegicus (Norway rat) ---------- -----
Mus musculus (house mouse) ---------- -----
Danio rerio (zebrafish) STANKQPEET STGNF 437

Fig. 1—Cont'd

B

	Human	Chimpanzee	Baboon-1	Baboon-2	Rhesus monkey	Whale	Cattle	Dog	Giant panda	Rat	Mouse	Zebrafish
Human	100	99	96	96	95	87	83	84	81	71	70	52
Chimpanzee		100	96	96	95	87	83	84	81	70	71	52
Baboon-1			100	100	99	87	85	83	81	70	69	52
Baboon-2				100	99	87	85	83	81	70	69	52
Rhesus monkey					100	87	84	82	81	70	69	53
Whale						100	89	87	82	71	71	52
Cattle							100	84	78	70	69	51
Dog								100	84	70	69	53
Giant panda									100	68	67	51
Rat										100	87	47
Mouse											100	47
Zebrafish												100

Fig. 1—Cont'd

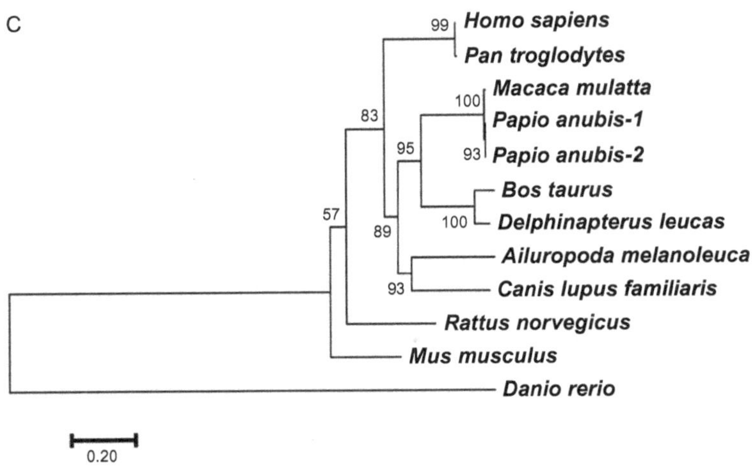

Fig. 1—Cont'd

is also observed in the medulla and occipital lobe.[9] The nucleus accumbens is an important reward center, regulating incentive salience, associative learning, and positive emotions.[2] About a third to quarter of XLAG patients have increased food seeking and further studies are thus needed to determine if the behavior of food seeking is regulated by high *GPR101* expression in nucleus accumbens.[5]

In the normal pituitary, GPR101 displays an age-dependent expression pattern with high expression in fetus and adolescence and low (or absent) expression in childhood and adult.[9] This result suggests that in the pituitary, GPR101 may regulate proliferation and maturity in fetal development and adolescence, while its physiological role may already be finished in adults.[9]

2.3.2 GPR101 expression in other species

Murine *Gpr101* mRNA is highly expressed in the hypothalamus, specifically in the arcuate nucleus, paraventricular nucleus, and supraoptic nucleus, suggesting a potential role of GPR101 in regulating multiple physiological functions such as hormone secretion and reproduction.[10] Murine *Gpr101* is also expressed in dopaminergic, serotonergic, and noradrenergic nuclei associated with mood modulation, motivational drive, and attentional state regulation.[10] In rat pituitary, GPR101 protein expression is only detected after birth, and sexual dimorphism of mRNA expression appears with 2- and 3-week females expressing significantly higher *Gpr101* mRNA levels.[9]

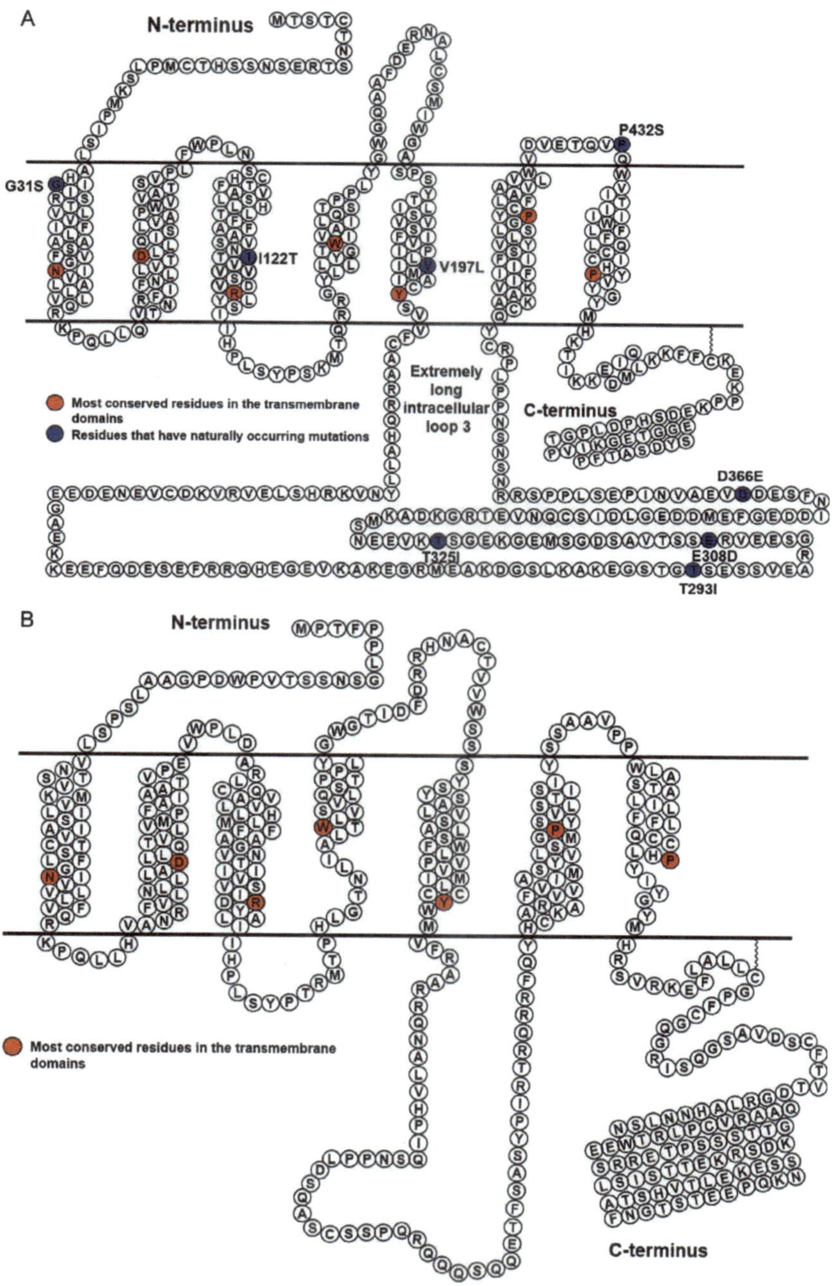

Fig. 2 Two-dimensional illustration of human (A) and zebrafish (B) GPR101. The most conserved residues in the transmembrane domains are shaded in *red*. Residues that have naturally occurring mutations or polymorphisms are shaded in *blue*.

In zebrafish, *gpr101* is expressed in a bimodal temporal expression pattern during embryonic development. In one-cell stage, a relatively high *gpr101* expression, which is probably maternal transcript, is detected by in situ hybridization, and the expression is decreasing from one-cell to eight-cell stage. The expression of *gpr101* reappears in bud stage, and strong expression is detected in the brain in 2-day postfertilization embryos.[9]

3. Physiology of GPR101

3.1 Energy balance

Murine GPR101 is highly expressed in the hypothalamus, where it might control energy homeostasis.[9] In the mouse brain, increased *Gpr101* mRNA expression is found in the posterior hypothalamus and amygdala after food deprivation.[11] However, in the obese *ob/ob* mice, *Gpr101* mRNA level is decreased in the posterior hypothalamus but remains unaltered in the amygdala after food deprivation.[11] In the hypothalamus, the subset of neurons that increases food intake expresses neuropeptide-Y (NPY) and Agouti-related peptide, while neurons that decreases food intake express proopiomelanocortin (POMC). In the arcuate nucleus, *Gpr101* mRNA is expressed in approximately 55% of neurons expressing *Pomc* but in only 5% of neurons expressing *Npy*, implying that if GPR101 has roles in energy regulation, it is more likely to regulate anorexigenic rather than orexigenic pathways.[11]

POMC-derived peptides include α-, β-, and γ-melanocyte stimulating hormone (α-, β-, and γ-MSH) and adrenocorticotropin. These peptide hormones display anorexigenic effects after binding to neural melanocortin-3 or melanocortin-4 receptor (MC3R or MC4R), two receptors that play important roles in regulating energy balance.[12,13] Further studies indicate that agonists and partial agonists of GPR101 stimulate hypothalamic α-, β-, and γ-MSH secretion, and inverse agonists and antagonists of GPR101 inhibit secretion of these peptide hormones.[14] This provides evidence that the potential pathway of GPR101 to regulate anorexigenic activities may act through the regulation of POMC signaling.

Leptin, the product of the *ob* gene, is known to regulate energy balance by upregulating *Pomc* and downregulating *Npy* gene expression.[15–17] Intraperitoneal injection of leptin to obese *ob/ob* mice can partially restore *Pomc* and *Npy* mRNA expression and cause a significant decrease in body weight; however, it does not alter *Gpr101* mRNA levels in the posterior hypothalamus and amygdala.[11] This suggests that GPR101-associated energy homeostasis is independent of leptin signaling in the posterior hypothalamus and amygdala.

3.2 Reproduction

Gpr101 mRNA is expressed in medial preoptic area (mPOA),[10] a major site where gonadotropin-releasing hormone (GnRH) neuron cell bodies are mainly localized,[18] and the arcuate nucleus, which is responsible for regulating 17β-estradiol (E_2) negative feedback as well as pulsatile release of GnRH.[19,20] GnRH is the master regulator of reproduction through regulating the hypothalamus–pituitary–gonad (HPG) axis, which ultimately results in steroid hormone secretion. GnRH-(1–5) is a GnRH metabolite that regulates HPG axis by exerting its actions through GPR101 and GPR173.[21–24] GnRH-(1–5) stimulates GnRH gene expression at the posttranscriptional level in the GnRH neuronal cell line (GT1–7 cell line) via a calcium/calmodulin-dependent protein kinase.[25]

Using an ovariectomized (OVX) rat model, Bauman et al. revealed an age-related *Gpr101* mRNA expression pattern. *Gpr101* mRNA expression is upregulated in the arcuate nucleus of older rats as compared to younger rats, while *Gpr101* mRNA expression within the pituitary and mPOA is unaltered during aging.[21] Moreover, an age-related downregulation of *Gpr101* expression in response to E_2 is also observed, as E_2 replacement increases *Gpr101* expression in younger OVX rats; however, older OVX rats treated with E_2 show decreased *Gpr101* expression. Although there is no classical estrogen-response element (ERE) in the promoter of *Gpr101*, it is possible that weak estrogen responsive motifs or ERE-independent signaling can regulate *Gpr101* expression.[21] Recent studies indicate that *Gpr173*, another receptor of GnRH-(1–5), is expressed in immortalized hypothalamic GnRH and kisspeptin neurons,[26] and Woitowich et al. showed that EP24.15, which cleaves GnRH to produce GnRH-(1–5), is also expressed in GnRH or kisspeptin neurons in mPOA or arcuate nucleus.[27] Consequently, further studies need to investigate whether *Gpr101* is also expressed in GnRH and/or kisspeptin neurons.

3.3 Cancer

Hyperactivation of epidermal growth factor receptor (EGFR) signaling is correlated with uncontrolled cell propagation, metastasis, and unresponsiveness to hormone therapy. Many GPCRs have been shown to transactivate EGFR by increasing membrane-bound matrix metalloprotease (MMP) activity.[28] MMP can cleave EGF ligands and induce ligand shedding into the extracellular space around EGFR. These cleaved EGF ligands can bind to the EGFR and other EGFR family members.[28]

Apart from regulation of the HPG axis, GnRH-(1–5) also has effects on peripheral tissues via GPR101. Using the Ishikawa cell line, an in vitro model of endometrial cancer, Cho-Clark et al. demonstrated that GnRH-(1–5) stimulates EGF release, EGFR phosphorylation, and cell migration. By using batimastat, an MMP inhibitor, it was shown that activation of GPR101 mobilizes membrane MMPs to release membrane-bound EGF. The released EGF can induce EGFR phosphorylation at the residues Y992, Y1045, and Y1068, thus activating the EGFR pathway as well as signaling cascades related to cellular migration[22] (Fig. 3A). Downregulation of GPR101 reverses GnRH-(1–5)-mediated EGFR activation and cellular migration, implicating that GnRH-(1–5) regulates EGF release and EGFR phosphorylation and cellular migration via GPR101 in the Ishikawa cell line.[22]

GnRH-(1–5) also stimulates GnRH-II mRNA expression in Ishikawa cell line[29] and GnRH-II can enhance cell motility and invasion, and promote EGFR activation.[30–33] It is possible that after binding to GPR101, GnRH-(1–5) stimulates downstream signaling that targets the GnRH-II systems to increase EGFR-dependent cell migration.[22]

DNA methylation is an epigenetic modification that is important in regulating gene expression, and alteration of a promoter's DNA methylation profile contributes to the initial formation and subsequent development of cancer.[34] Colorectal cancer is one of the most frequent malignancies, therefore methylated DNA from colorectal tumor tissue and normal colonic mucosa was analyzed to identify alterations in promoter methylation.[35] From studies on 77 colorectal cancer patients, *GPR101* was identified as one of nine novel genes with promoter hypermethylation. Methylation of *GPR101* promoter is found in 40% of cases. Kober et al. further showed that in stage IV male colorectal cancer patients, *GPR101* methylation correlates with longer time to progression (defined as a time from surgery to local recurrence, metastasis growth, or new metastases occurrence), with survival benefit.[35] This study is based on male patients, and further investigations on female patients are required.

4. Pharmacology of GPR101

4.1 Ligand binding and receptor activation

After alignment of GPR101 sequence with other Family A GPCR sequences, GPR101 falls into a class of orphan GPCRs whose ligands are close relatives of melatonin receptors and opsins.[36] However, it is impossible

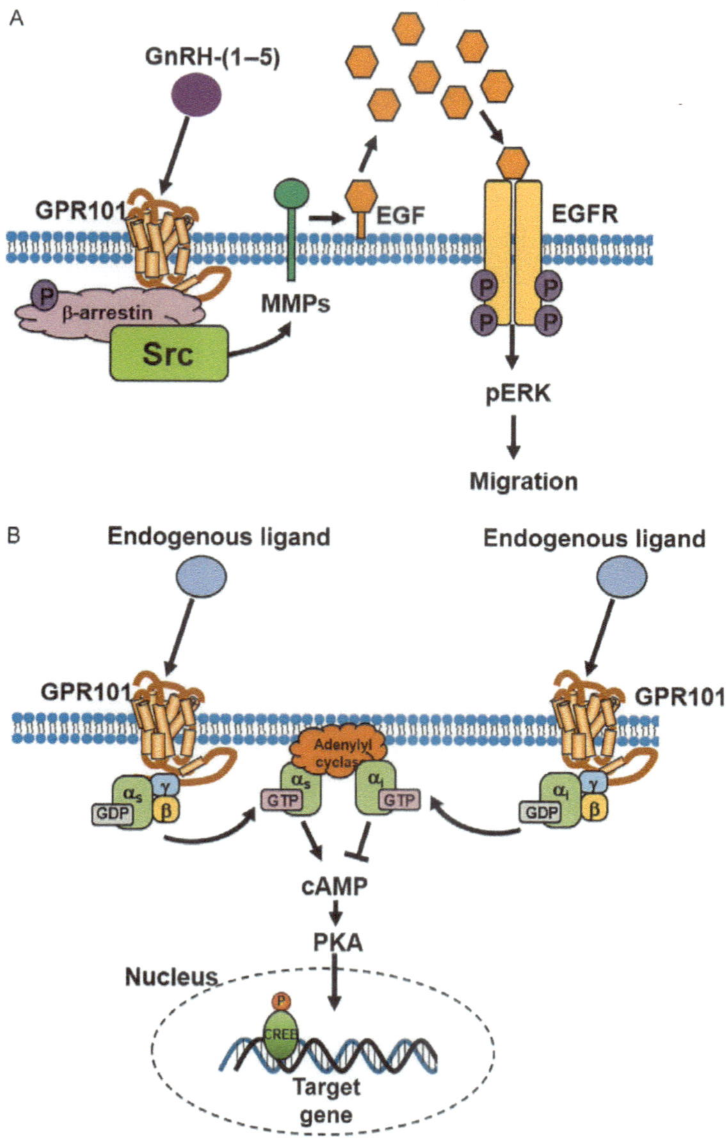

Fig. 3 Mechanisms of GPR101-mediated physiological functions. (A) After GnRH-1–5 binding, GPR101 possibly promotes matrix metalloprotease (MMP) activity via β-arrestin-Src signaling. Membrane-bound EGF is cleaved by MMP and EGF released can induces EGFR phosphorylation, thus activating EGFR pathway and signaling cascades related to cell migration. (B) GPR101 may be coupled to both $G\alpha_s$ and $G\alpha_i$ proteins to stimulate or inhibit intracellular cAMP accumulation, thus regulating gene transcription.

to predict GPR101 endogenous ligands based on sequence similarity, as the similarity between these receptors is low. GPR101 sequence is somewhat similar to GPR161, another orphan receptor, and with histamine H2 receptor, but some hallmarks of biogenic amine receptors, such as the aspartate in transmembrane domain three, are absent in GPR101. This suggests that the biogenic amine receptor ligands are unlikely to act as GPR101 endogenous ligands.[10] In an attempt to identify ligands that stimulate or inhibit GPR101, Bates et al. incubated panels of small molecules, peptides, and bioactive lipids with mammalian and yeast systems. However, no selectively active compound was identified as GPR101 ligand.[10]

Using GN11 cells, a neuronal cell line with premigratory GnRH secretion, Larco et al. demonstrated that GnRH-(1–5) significantly inhibits GN11 cell migration.[37] A panel of 74 orphan GPCRs was used to determine whether GnRH-(1–5) regulates GN11 migration through GPCR activation. The results showed that GnRH-(1–5) can specifically activate GPR101, GPR119, GPR148, and GPR173, though only GPR173 mediates the inhibition of migration.[37] Cho-Clark et al. showed that after binding to GPR101, GnRH-(1–5) increases MMP-9 activity, resulting in rapid EGF release, EGFR phosphorylation, and subsequent promotion of Ishikawa cell invasion[22,23] (Fig. 3A).

4.2 Signaling pathways

As no endogenous GPR101 ligand has been identified, it is extremely difficult to characterize GPR101 G protein-coupling specificity by using ligand stimulation. Therefore, a knowledge-restricted hidden Markov model is developed to predict GPR101 G protein-coupling specificity.[38] This model hypothesizes that a contiguous sequence according to the concatenated four cytosolic domains (three intracellular loops and the cytosolic C-tail) of GPCRs could act as a discriminator to predict the G protein-coupling specificity.[38] G protein-coupling selectivity is predicted based on the previous hypothesis that G protein selectivity is defined by the intracellular conformation that is regulated by the interaction between intracellular regions.[39] Within an error rate less than 1%, both mouse and human GPR101 are predicted as $G\alpha_s$-coupling GPCRs[10] (Fig. 3B).

Basal signaling of GPR101 was evaluated in different cell lines including HEK293 cells (human embryonic kidney cells), GH3 cells (rat GH3 cell line), CHO (Chinese hamster ovary) cells, and yeast cells, respectively. Expression plasmids that encode human GPR101 were transiently transfected into HEK293 cells containing stably expressed cAMP response element-luciferase reporter (CRE-luciferase reporter) gene, which can be activated by $G\alpha_s$-coupled GPCRs. Transfection of human GPR101 elicits an elevated, agonist-independent, luciferase response, and intracellular cAMP level that is proportional to the amount of transfected human GPR101.[10] In GH3 cells, similar constitutive activation of Gs–cAMP pathway was observed.[4]

The chimeric Gpa1 proteins of Gpa1-s5, Gpa1-o5, and Gpa1-q5 represent $G\alpha_s$, $G\alpha_i$, and $G\alpha_q$ subunits, respectively.[40] In yeast, strains individually expressing these distinct chimeric Gα proteins and either a GPR101 expression vector or empty vector were tested. Expression of GPR101 in Gpa1-s5-containing cells, but not in Gpa1-o5- or Gpa1-q5-containing cells, elicits an elevated β-galactosidase response, consistent with the G protein-coupling preferences for $G\alpha_s$ and constitutive activation of Gs–cAMP pathway.[10]

However, in CHO cells, no constitutive activation of Gs–cAMP was observed for GPR101.[41] Instead, GPR101 inhibits CRE-mediated gene expression stimulated by forskolin,[41] suggesting the possibility that GPR101 may be coupled to both $G\alpha_s$ and $G\alpha_i$ proteins, as previously reported for M2 acetylcholine receptor.[42]

GnRH-(1–5) can induce β-arrestin binding to GPR101, indicating receptor activation.[24] Treatment of Ishikawa cells with GnRH-(1–5) results in EGF release, EGFR phosphorylation, and cell invasion without significant changes in cAMP levels, indicating that GnRH-(1–5) activates GPR101 in a cAMP-independent mechanism.[23] After ligand binding, GPCRs can trigger G protein-dependent and/or β-arrestin-dependent intracellular signaling pathways.[43] β-arrestin can directly interact with Src family tyrosine kinases and GPCRs possibly promote MMPs activity via β-arrestin-Src signaling.[44,45] It remains to be further investigated whether, after binding to GPR101, GnRH-(1–5) selectively stimulates β-arrestin-dependent EGFR activation.

5. Pathophysiology of GPR101

5.1 *GPR101* duplication

Gigantism due to GH oversecretion in childhood can be due to different genetic causes. *AIP* mutations (29%–41%) are the most common genetic cause of gigantism.[2] Gigantism resulting from *AIP* mutations typically exhibits clinical traits in the second decade of life. By contrast, XLAG results in early-onset gigantism due to germline or somatic microduplication of

GPR101.[6] In two reports with large numbers of cases of pituitary gigantism, XLAG represents 7.8% and 10% of the cases, respectively.[6,46] Up to now, 33 XLAG patients have been identified, with 9 males and 24 females, 26 patients with germline *GPR101* duplication, 6 patients with somatic *GPR101* duplication, and 1 patient undermined.[2]

Patients with XLAG usually display excessive growth before the age of one, even as young as 2–3 months in some patients, though born with normal body sizes.[2] Growth in prenatal stage is mediated mainly by IGF-2, while growth in neonatal life is regulated by the GHRH/SST-GH-IGF-1 axis.[5] It may be speculated that IGF-2 secretion is independent of *GPR101*, and abnormal growth during neonatal life is induced by increased circulating GH due to the pituitary lesion.[2,3] Although histopathological observation reveals significant hyperplasia of somatotrophs, lactotrophs, and mammosomatotrophs (cells expressing both GH and prolactin) in some XLAG patients,[47,48] most XLAG patients harbor pituitary adenoma.[2] In three patients, pituitary adenoma was surrounded by hyperplasia, implying that hyperplasia to adenoma progression occurs in these XLAG patients.[2] Increased circulating GHRH, IGF-1 and prolactin levels are also observed in some XLAG patients and GHRH upregulation may further increase the already elevated basal GH and prolactin levels and the hyperplasia to adenoma progression.[3,49]

5.2 *GPR101* mutations

When *GPR101* was sequenced in 248 patients with sporadic acromegaly, Trivellin et al. identified a *GPR101* mutation (c.924G > C, E308D) in 11 patients.[4] Of the 11 mutation carriers, 8 patients carry somatic mutation and 3 patients have germline mutation.[4] The E308 residue is located in the remarkably long ICL3 in GPR101, connecting transmembrane domains 5 and 6. When E308D GPR101 was transfected into GH3 cells, increased cell proliferation and GH oversecretion are observed.[4] In another study, Lacovazzo et al. investigated 579 acromegaly patients, identifying 4 patients (0.69%) harboring germline *GPR101* E308D mutation, although there is no family history of pituitary adenoma. In this cohort, acromegaly patients do not have an elevated prevalence of E308D *GPR101* mutation when compared to public database, Exome Aggregation Consortium (ExAC) database.[6] Ferraù et al. investigated 215 acromegaly patients and Matsumoto et al. analyzed 61 patients with somatotropinoma that may attribute to acromegaly.[50,51] None of these patients carry *GPR101* E308D mutation.[50,51]

GPR101 E308D mutation also results in a small fraction of pituitary adenomas that are not associated with GH oversecretion.[52] After investigation of germline *GPR101* mutations in 766 patients with sporadic pituitary adenomas, the result demonstrated that pituitary adenoma patients have a higher allele frequency of *GPR101* mutation than that of unaffected individuals included in the ExAC database.[52] Six patients carrying germline *GPR101* E308D mutation were identified. Apart from three patients who have adult-onset acromegaly due to pure GH-secreting adenoma, two patients with mutation have adrenocorticotropin-secreting adenomas and one patient has panhypopituitarism.[52]

Other *GPR101* mutations have also been identified in patients with acromegaly or other endocrinological diseases.[52] Lecoq et al. reported six patients carrying *GPR101* mutations, including one acromegaly patient with a *GPR101* mutation (c.1098C > A, D366E), two patients with prolactin-secreting adenomas harboring *GPR101* mutations (c.974C > T, T325I; c.1294C > T, P432S), one Cushing's disease patient with a *GPR101* mutation (c.365 T > Y, I122T), one patient with TSH- and GH-co-secreting adenoma harboring a *GPR101* silent mutation (c.354C > T), and one nonfunctioning pituitary adenoma patient carrying *GPR101* silent mutation (c.1074C > T).[52] *GPR101* is highly expressed in POMC neurons and *GPR101* E308D and I122T mutations have been shown to associate with ACTH oversecretion.[11,52] Trivellin et al. sequenced DNA from peripheral blood and pituitary tumor in 36 patients with Cushing's disease, identifying a rare *GPR101* mutation (c.91G > A, G31S) in one patient with Cushing's disease. Compared to wild type, overexpression of *GPR101* G31S does not increase cell proliferation in rat and mouse pituitary cell lines (GH3 and AtT-20 cells), suggesting that *GPR101* G31S mutant is unlikely to be involved in the pathogenesis of Cushing's disease.[53]

Given the phenotypes of XLAG induced by *GPR101* duplication and acromegaly resulting from *GPR101* mutations, Castinetti et al. sequenced germline *GPR101* in 41 patients with congenital isolated GH deficiency in order to determine whether loss-of-function mutations in *GPR101* were involved.[54] One novel *GPR101* mutation c.589G > T (V197 L) was identified in one patient. Compared to wild type, this mutant does not have any significant difference in stimulating GH secretion and cAMP response.[54] Further study is needed to assess the clinical significance of *GPR101* V197L mutation.

6. Conclusion

XLAG, a recently described early-onset gigantism that is characterized by oversecretion of GH, accounts for a significant proportion of pituitary gigantism cases. XLAG is attributed to an Xq26.3 genomic duplication that

GPR101 is invariably involved. GPR101 is mainly expressed in brain and pituitary, and GPR101 is involved in regulation of energy balance, reproduction, and cancer cell migration. GPR101 was first identified as an orphan receptor. An GnRH metabolite, GnRH-(1–5), is reported to be able to activate GPR101 and induce EGF release, EGFR phosphorylation, and cell invasion. However, intracellular cAMP levels are not altered after GnRH-(1–5) binding, suggesting that GnRH-(1–5) activates GPR101 in cAMP-independent manner. Whether GnRH-(1–5) activates GPR101 and stimulates EGFR phosphorylation in a β-arrestin-dependent manner requires further studies. Apart from the causative role of *GPR101* duplication that leads to XLAG, *GPR101* mutations, especially *GPR101* E308D, are identified in acromegaly patients. *GPR101* mutations are also found in patients with other pituitary tumors without GH oversecretion. Further studies are needed to identify the endogenous ligands of GPR101 and clarify how *GPR101* duplication and/or *GPR101* mutation lead to GH oversecretion in patients with pituitary tumor.

Acknowledgment

Z.S.H. was supported by a fellowship from China Scholarship Council of the People's Republic of China.

References

1. Veldhuis JD, Iranmanesh A, Erickson D, Roelfsema F, Bowers CY. Lifetime regulation of growth hormone (GH) secretion. In: *Handbook of Neuroendocrinology*. Elsevier; 2012:237–257.
2. Trivellin G, Hernández-Ramírez C, Swan J, Stratakis CA. An orphan G protein-coupled receptor causes human gigantism and/or acromegaly: molecular biology and clinical correlations. *Best Pract Res Clin Endocrinol Metab*. 2018;32:125–140.
3. Iacovazzo D, Korbonits M. Gigantism: X-linked acrogigantism and *GPR101* mutations. *Growth Horm IGF Res*. 2016;30:64–69.
4. Trivellin G, Daly AF, Faucz FR, et al. Gigantism and acromegaly due to Xq26 microduplications and *GPR101* mutation. *N Engl J Med*. 2014;371:2363–2374.
5. Beckers A, Lodish MB, Trivellin G, et al. X-linked acrogigantism syndrome: clinical profile and therapeutic responses. *Endocr Relat Cancer*. 2015;22:353–367.
6. Iacovazzo D, Caswell R, Bunce B, et al. Germline or somatic *GPR101* duplication leads to X-linked acrogigantism: a clinico-pathological and genetic study. *Acta Neuropathol Commun*. 2016;4:56.
7. Kakarala KK, Jamil K. Sequence-structure based phylogeny of GPCR class A rhodopsin receptors. *Mol Phylogenet Evol*. 2014;74:66–96.
8. Lee DK, Nguyen T, Lynch KR, et al. Discovery and mapping of ten novel G protein-coupled receptor genes. *Gene*. 2001;275:83–91.
9. Trivellin G, Bjelobaba I, Daly AF, et al. Characterization of *GPR101* transcript structure and expression patterns. *J Mol Endocrinol*. 2016;57:97–111.

10. Bates B, Zhang L, Nawoschik S, et al. Characterization of *Gpr101* expression and G-protein coupling selectivity. *Brain Res*. 2006;1087:1–14.
11. Nilaweera KN, Ozanne D, Wilson D, Mercer JG, Morgan PJ, Barrett P. G protein-coupled receptor 101 mRNA expression in the mouse brain: altered expression in the posterior hypothalamus and amygdala by energetic challenges. *J Neuroendocrinol*. 2007;19:34–45.
12. Tao YX. The melanocortin-4 receptor: physiology, pharmacology, and pathophysiology. *Endocr Rev*. 2010;31:506–543.
13. Tao YX. Mutations in the melanocortin-3 receptor (MC3R) gene: impact on human obesity or adiposity. *Curr Opin Investig Drugs*. 2010;11:1092–1096.
14. Bagnol D. Methods of using GPR101 receptors to identify modulators of hypothalamic proopiomelanocortin (POMC)-derived biologically active peptide secretion. Google Patents; 2012. WO2007/142979.
15. Kalra SP, Dube MG, Pu S, Xu B, Horvath TL, Kalra PS. Interacting appetite-regulating pathways in the hypothalamic regulation of body weight. *Endocr Rev*. 1999;20:68–100.
16. Cheung CC, Clifton DK, Steiner RA. Proopiomelanocortin neurons are direct targets for leptin in the hypothalamus. *Endocrinology*. 1997;138:4489–4492.
17. Mercer JG, Hoggard N, Williams LM, et al. Coexpression of leptin receptor and preproneuropeptide Y mRNA in arcuate nucleus of mouse hypothalamus. *J Neuroendocrinol*. 1996;8:733–735.
18. Powers B, Valenstein ES. Sexual receptivity: facilitation by medial preoptic lesions in female rats. *Science*. 1972;175:1003–1005.
19. Maeda K-I, Adachi S, Inoue K, Ohkura S, Tsukamura H. Metastin/kisspeptin and control of estrous cycle in rats. *Rev Endocr Metab Disord*. 2007;8:21–29.
20. Goodman RL, Hileman SM, Nestor CC, et al. Kisspeptin, neurokinin B, and dynorphin act in the arcuate nucleus to control activity of the GnRH pulse generator in ewes. *Endocrinology*. 2013;154:4259–4269.
21. Bauman BM, Yin W, Gore AC, Wu TJ. Regulation of GnRH-(1-5) signaling genes by estradiol is age-dependent. *Front Endocrinol (Lausanne)*. 2017;8:282.
22. Cho-Clark M, Larco DO, Semsarzadeh NN, Vasta F, Mani SK, Wu TJ. GnRH-(1-5) transactivates EGFR in Ishikawa human endometrial cells via an orphan G protein-coupled receptor. *Mol Endocrinol*. 2014;28:80–98.
23. Cho-Clark M, Larco DO, Zahn BR, Mani SK, Wu TJ. GnRH-(1-5) activates matrix metallopeptidase-9 to release epidermal growth factor and promote cellular invasion. *Mol Cell Endocrinol*. 2015;415:114–125.
24. Larco DO, Semsarzadeh NN, Cho-Clark M, Mani SK, Wu TJ. The novel actions of the metabolite GnRH-(1-5) are mediated by a G protein-coupled receptor. *Front Endocrinol (Lausanne)*. 2013;4:83.
25. Wu TJ, Mani SK, Glucksman MJ, Roberts JL. Stimulation of luteinizing hormone-releasing hormone (LHRH) gene expression in GT1-7 cells by its metabolite, LHRH-(1-5). *Endocrinology*. 2005;146:280–286.
26. Treen AK, Luo V, Belsham DD. Phoenixin activates immortalized GnRH and kisspeptin neurons through the novel receptor GPR173. *Mol Endocrinol*. 2016;30:872–888.
27. Woitowich NC, Philibert KD, Leitermann RJ, Wungjiranirun M, Urban JH, Glucksman MJ. EP24.15 as a potential regulator of kisspeptin within the neuroendocrine hypothalamus. *Endocrinology*. 2015;157:820–830.
28. George AJ, Hannan RD, Thomas WG. Unravelling the molecular complexity of GPCR-mediated EGFR transactivation using functional genomics approaches. *FEBS J*. 2013;280:5258–5268.
29. Baldwin EL, Wegorzewska IN, Flora M, Wu TJ. Regulation of type II luteinizing hormone-releasing hormone (LHRH-II) gene expression by the processed peptide of

LHRH-I, LHRH-(1-5) in endometrial cells. *Exp Biol Med (Maywood)*. 2007;232: 146–155.
30. Poon SL, Hammond GT, Leung PCK. Epidermal growth factor-induced GnRH-II synthesis contributes to ovarian cancer cell invasion. *Mol Endocrinol*. 2009;23:1646–1656.
31. Poon SL, Lau MT, Hammond GL, Leung PCK. Gonadotropin-releasing hormone-II increases membrane type I metalloproteinase production via β-catenin signaling in ovarian cancer cells. *Endocrinology*. 2011;152:764–772.
32. Poon SL, Klausen C, Hammond GL, Leung PCK. 37-kDa laminin receptor precursor mediates GnRH-II-induced MMP-2 expression and invasiveness in ovarian cancer cells. *Mol Endocrinol*. 2011;25:327–338.
33. Wu HM, Wang HS, Huang HY, et al. Gonadotropin-releasing hormone type II (GnRH-II) agonist regulates the invasiveness of endometrial cancer cells through the GnRH-I receptor and mitogen-activated protein kinase (MAPK)-dependent activation of matrix metalloproteinase (MMP)-2. *BMC Cancer*. 2013;13:300.
34. Jones PA. DNA methylation and cancer. *Cancer Res*. 1986;46:461–466.
35. Kober P, Bujko M, Olędzki J, Tysarowski A, Siedlecki JA. Methyl-CpG binding column-based identification of nine genes hypermethylated in colorectal cancer. *Mol Carcinog*. 2011;50:846–856.
36. Vassilatis DK, Hohmann JG, Zeng H, et al. The G protein-coupled receptor repertoires of human and mouse. *Proc Natl Acad Sci USA*. 2003;100:4903–4908.
37. Larco DO, Cho-Clark M, Mani SK, Wu TJ. The metabolite GnRH-(1-5) inhibits the migration of immortalized GnRH neurons. *Endocrinology*. 2013;154:783–795.
38. Sreekumar KR, Huang Y, Pausch MH, Gulukota K. Predicting GPCR-G-protein coupling using hidden Markov models. *Bioinformatics*. 2004;20:3490–3499.
39. Wong SKF. G protein selectivity is regulated by multiple intracellular regions of GPCRs. *Neurosignals*. 2003;12:1–12.
40. Erlenbach I, Kostenis E, Schmidt C, et al. Single amino acid substitutions and deletions that alter the G protein coupling properties of the V2 vasopressin receptor identified in yeast by receptor random mutagenesis. *J Biol Chem*. 2001;276:29382–29392.
41. Martin AL, Steurer MA, Aronstam RS. Constitutive activity among orphan class-A G protein coupled receptors. *PLoS One*. 2015;10, e0138463.
42. Michal P, Lysíková M, Tuček S. Dual effects of muscarinic M2 acetylcholine receptors on the synthesis of cyclic AMP in CHO cells: dependence on time, receptor density and receptor agonists. *Br J Pharmacol*. 2001;132:1217–1228.
43. Tao YX, Conn PM. Chaperoning G protein-coupled receptors: from cell biology to therapeutics. *Endocr Rev*. 2014;35:602–647.
44. Luttrell LM, Lefkowitz RJ. The role of β-arrestins in the termination and transduction of G protein-coupled receptor signals. *J Cell Sci*. 2002;115:455–465.
45. New DC, Wong YH. Molecular mechanisms mediating the G protein-coupled receptor regulation of cell cycle progression. *J Mol Signal*. 2007;2:2.
46. Rostomyan L, Daly AF, Petrossians P, et al. Clinical and genetic characterization of pituitary gigantism: an international collaborative study in 208 patients. *Endocr Relat Cancer*. 2015;22:745–757.
47. Rodd C, Millette M, Iacovazzo D, et al. Somatic *GPR101* duplication causing X-linked acrogigantism (XLAG)—diagnosis and management. *J Clin Endocrinol Metab*. 2016;101:1927–1930.
48. Moran A, Asa SL, Kovacs K, et al. Gigantism due to pituitary mammosomatotroph hyperplasia. *N Engl J Med*. 1990;323:322–327.
49. Daly AF, Lysy P, Defilles C, et al. Growth hormone releasing hormone excess and blockade in X-LAG syndrome. *Endocr Relat Cancer*. 2016;23:161–170.
50. Ferrau F, Romeo PD, Puglisi S, et al. Analysis of *GPR101* and *AIP* genes mutations in acromegaly: a multicentric study. *Endocrine*. 2016;54:762–767.

51. Matsumoto R, Izawa M, Fukuoka H, et al. Genetic and clinical characteristics of Japanese patients with sporadic somatotropinoma. *Endocr J.* 2016;63:953–963.
52. Lecoq A-L, Bouligand J, Hage M, et al. Very low frequency of germline *GPR101* genetic variation and no biallelic defects with *AIP* in a large cohort of patients with sporadic pituitary adenomas. *Eur J Endocrinol.* 2016;174:523–530.
53. Trivellin G, Correa RR, Batsis M, et al. Screening for *GPR101* defects in pediatric pituitary corticotropinomas. *Endocr Relat Cancer.* 2016;23:357–365.
54. Castinetti F, Daly AF, Stratakis CA, et al. *GPR101* mutations are not a frequent cause of congenital isolated growth hormone deficiency. *Horm Metab Res.* 2016;48:389–393.

CHAPTER FOUR

Diseases caused by mutations in luteinizing hormone/chorionic gonadotropin receptor

Jie Qiao*, Bing Han
Department of Endocrinology, Shanghai Ninth People's Hospital, Shanghai Jiao Tong University School of Medicine, Shanghai, People's Republic of China
*Corresponding author: e-mail address: qiaoj2001@126.com

Contents

Abstract

Accumulating evidence showed that the luteinizing hormone/chorionic gonadotropin receptor (LHCGR) is an essential regulator of sexual development and reproduction from zebrafish to human. Activating and inactivating mutations of *LHCGR* gene have been identified from patients of different phenotypes. Familial male-limited precocious puberty, Leydig cell hypoplasia, and empty follicle syndrome are caused by *LHCGR* mutations. More than 50 mutations have been reported from subjects of different ethnic backgrounds. Functional analyses of the mutant LHCGR revealed multiple defects, including cell surface expression, ligand binding, and signaling. The difference of the two native ligands and signaling pathway activated by LHCGR are illustrated. Potential therapeutic implications from the analyses of the naturally occurring *LHCGR* mutations, such as pharmacological chaperones, are highlighted.

Progress in Molecular Biology and Translational Science, Volume 161
ISSN 1877-1173
https://doi.org/10.1016/bs.pmbts.2018.09.007

Abbreviations

DAG	diacylglycerol
DHT	dihydrotestosterone
ECD	extracellular domain
EFS	empty follicle syndrome
FMPP	familial male-limited precocious puberty
GPCR	G protein-coupled receptor
hCG	human chorionic gonadotropin
ICL	intracellular loop
IP3	inositol trisphosphate
IVF	in vitro fertilization
LCH	Leydig cell hypoplasia
LH	luteinizing hormone
LHCGR	luteinizing hormone/chorionic gonadotropin receptor
LRR	leucine-rich repeat
PCOS	polycystic ovary syndrome
PKC	protein kinase C
PLC/IP	phospholipase C/inositol phosphate
T	testosterone
TM	transmembrane

1. Introduction

The luteinizing hormone/chorionic gonadotropin receptor (LHCGR) plays a critical role in human sexual development and reproduction. The receptor is shared between the placental glycoprotein hormone human chorionic gonadotropin (hCG) and luteinizing hormone (LH). It is well known that hCG exerts its effect during early embryogenesis and induces the maturation of Leydig cells, whereas LH is required for the induction of ovulation in females and for sex steroid hormone production in both males and females.

In this chapter, after reviewing the physiology of LHCGR, we will look at the main clinical features of gain-of-function and loss-of-function mutations of the *LHCGR* gene, as well as the biological, cellular, and molecular tools used to investigate the LHCGR. Numerous *LHCGR* mutations have been described over the past 26 years. We will consider the different clinical features by examining the genotype and phenotype of the LHCGR-related diseases. Polymorphic variants identified within *LHCGR* and their contribution in determining clinical phenotypes are also reviewed. Lastly, we will address the role of LHCGR in assisted reproduction, as well as potential therapeutic effect of small-molecule allosteric agonist that functionally rescues intracellularly retained mutant LHCGRs.

2. Physiology and structure of LHCGR

LHCGR signaling plays an essential role in reproduction through the transduction of the signal of the mid-cycle LH surge, leading to ovulation and subsequent maintenance of progesterone production by the corpus luteum. Consistent with its role in female reproduction, LHCGR is expressed by multiple cell types in the ovary, including theca, luteal, interstitial, and differentiated granulosa cells.[1] During the mid-follicular phase, LH stimulates androgen production in theca cells. These androgens are aromatized in granulosa cells to estrogen. Prolonged, sustained high estrogen levels generate positive feedback on the pituitary gland and the subsequent LH surge that results in ovulation.[2] After ovulation, LH promotes progesterone production, supporting development of the corpus luteum.[3] LHCGR is downregulated transiently after the preovulatory LH surge, reaches a maximum at mid-luteal phase, and decreases with corpus luteum regression.

During pregnancy, hCG, as the second ligand for LHCGR, plays an important role in sustaining progesterone synthesis.[4,5] It is believed that hCG plays an important role in placentation through activities such as maintaining angiogenesis of the uterine vasculature, promoting differentiation of cytotrophoblasts into syncytiotrophoblasts, trophoblast invasion, and maternal recognition of pregnancy through LHCGR expressed in cytotrophoblasts and intermediate trophoblasts.[6,7] In addition to supporting implantation, trophoblast invasion, and placentation, hCG is also implicated in uterine growth control as well as growth and differentiation of fetal organs. In male fetuses, testosterone (T) production by fetal Leydig cells, which is under the control of placental hCG, begins midway through the first trimester, at approximately 7 weeks of gestation, reaching optimal level between 8 and 16 of gestation.[8] Thereafter, T production continues under the control of fetal LH at a lower level.[9]

As glycoprotein hormones, LH and hCG share a common 92-amino acid α-subunit, with the unique functions and receptor-binding capacity of each of these hormones stemming from differences among the β subunits.[10] CG is produced in primates, equines, and in man, whereas LH is present in all species.[11–13] HCG is more potent than LH due to its higher receptor-binding affinity and a longer circulatory half-life.[14,15] In total, more than 30 hLH isoforms and 15 variants of hCG have been identified from serum and urine samples.[16]

Masculinization of internal and external genitalia is androgen dependent, mediated by T and its 5α-reduced product dihydrotestosterone (DHT). Two different *LHCGR* knockout murine models (LHRKO and LuRKO) with similar phenotypes were constructed by two independent groups.[17,18] By targeted disruption of exon 11 of the *Lhcgr* gene, Zhang et al.[17] observed that testicular growth and descent, external genitalia, and accessory sex organ maturation are impaired in knockout males postnatally, although homozygous mice are born phenotypically normal, with testes and genital structures indistinguishable from their wild-type littermates. Moreover, spermatogenesis is arrested at the round spermatid stage. The ovaries of knockout females are smaller, lacking preovulatory follicles or corpora lutea. Reduced gonadal sex hormone production is found in each sex, which is also reflected by the suppressed accessory sex organ weights and elevated gonadotropin levels. Therefore, the evidence from this knockout mouse model supports that intrauterine sex differentiation is independent of LH action, but it has a crucial role postnatally for attaining sexual maturity in each sex.[17,19] Lei et al.[18] created LHRKO model by targeted deletion of the proximal part of the *LHCGR* promoter and exon 1. In female knockout mice, marked growth reduction of uterus, decreased thickness of endometrial and myometrial layers, and height of luminal epithelium and vascular space are observed. Twenty-one-day estradiol and progesterone replacement therapy does not normalize the decrease in the number of endometrial glands and several genes that either decreased or increased in expression.[18]

3. The *LHCGR* gene and LHCGR protein

The human *LHCGR* gene, located on chromosome 2p21, contains 11 exons. The first 10 exons encode the extracellular domain (ECD), while the last exon encodes a small portion of the ECD, the transmembrane domain, and the cytoplasmic C-terminal domain.[1,8,20] It is mainly expressed in testicular Leydig cells, ovarian theca cells, and differentiated granulosa cells.[21] A cryptic exon derived from potential splicing sites in intron 6 is found to cause intron retention and producing a cryptic exon (exon 6A).[22] The intron 6 of *LHCGR* contains a 2.7-kb-long genomic region which is only present in primates and humans.[22] Two alternative splicing donor sites (GT) have been identified, which, together with the 3′-acceptor site (AG), give rise to a 159-bp (short) or a 207-bp (long) internal exon.[23] Both short and long mRNAs contain stop codons triggering nonsense-mediated mRNA decay in *LHCGR*. When exon 6A is included, the

LHCGR protein will truncate to a protein of 209 amino acids. Functional analysis indicated the truncated protein is mainly expressed in cytoplasm and has no ligand-binding ability. Cotransfection of wild-type and truncated LHCGR in HEK293 cells does not reveal any dominant negative effect of LHCGR-6A.[23]

LHCGR belongs to a member of the family A or rhodopsin-like family of G protein-coupled receptors (GPCRs). The structure of GPCRs was constructed and modified based on the crystal structure of rhodopsin. It is composed of seven transmembrane (TM) helices connected by three extracellular and three intracellular loops and a large N-terminal ECD which is responsible for high-affinity binding of hCG or LH. The ECD of LHCGR contains a series of leucine-rich repeats (LRRs), which forms a horseshoe structure and serves as binding site for ligand.[1] Considering the similarity of crystal structures of family A GPCRs, we could predict the structure of hormone binding area of LHCGR by homology modeling using the structure of the hormone binding domain of human FSHR as the template.[24] Previously, based on site-directed mutagenesis studies, LRRs 2–8 are found to be crucial for high-affinity binding of LH and CG.[25–28] The polypeptide chains of the mature human LHCGR are predicted to be composed of 675 residues, with molecular mass approximately 75 kDa, including a 335-amino acid ECD with 6 putative glycosylation sites, a 267-amino acid region that displays 7 transmembrane segments, and a 72-amino acid C-terminal intracellular domain. The mature form of the LHCGR present at the cell surface is a glycoprotein with an apparent molecular mass of 85–95 kDa monomer as well as dimeric and oligomeric forms.[29,30]

With the new crystal structure of FSHR reported by Jiang et al. in 2012, we could advance our understanding of characteristics of LHCGR.[31] Similar to other GPCRs, a second hormone binding site was identified at the C-terminal part of the LHCGR hinge region.[32,33] Hence, there are two separate binding sites of the ligand at the extracellular region of the receptor, one at the LRRs and the other at the hinge region around Tyr331.[34]

4. Signaling pathways activated by the LHCGR

The effects of LHCGR on Leydig cells and granulosa cells are mediated primarily by the activation of the Gs/adenylyl cyclase/cAMP/PKA pathway, especially acting as a significant effector of downstream events that lead to ovulation and steroid biosynthesis. There is a growing body of evidence suggesting that LHCGR can independently activate two

G protein-dependent signaling pathways: adenylyl cyclase and the phospholipase C/inositol phosphate (PLC/IP) signaling pathway. Inositol trisphosphate (IP3) binds calcium channels located in the endoplasmic reticulum, resulting in Ca^{2+} release in a hormone concentration-dependent manner.[35] Gudermann et al.[36] observed that PLC-based signaling only occurs during the preovulatory LH surge and during pregnancy. Although both the cAMP/PKA and the PLC/Ca^{2+} pathways are activated simultaneously, the hCG half-maximal effective dose (EC_{50}) is 20-fold higher for Ca^{2+} mobilization than that needed for cAMP recruitment, and is independent of receptor density, demonstrating that LHCGR carries dual signaling potential.[37]

AKT and extracellular signal-regulated protein kinases 1 and 2 (ERK1/2) have been identified as major downstream players in LHCGR-mediated signaling, with putative roles in cell proliferation, differentiation, and survival.[38,39] Recent studies demonstrated that intracellular signaling, downstream events, and cell fate are differentially mediated by LH and hCG. It is suggested that LH binding has a greater impact on phosphorylation of AKT and ERK1/2 than hCG does, whereas binding of hCG generates a greater intracellular accumulation of cAMP than does LH.[40–42] LH action is preferentially exerted through kinases, ERK1/2 and AKT, resulting in irreplaceable proliferative/antiapoptotic signals and partial agonism on progesterone production in vitro. In contrast, hCG displays notable cAMP/PKA-mediated steroidogenic and proapoptotic potential, which is masked by estrogen action in vivo.[35]

Dufau et al.[43,44] found the basal promoter activity of the *LHCGR* to be within the first 176 bp of the 5′-flanking region, containing consensus sequences for two Sp1 sites, three AP-2 sites, and one estrogen receptor response element half-site (EREhs). Three nuclear orphan receptors (EAR2, EAR3/COUP-TF1, and TR4) that bind to this EREhs were identified. The results in transgenic mice indicated that LHCGR is mainly expressed in adult Leydig cells, but not in fetal Leydig cells or in ovarian cells. Besides, FSH-dependent induction of LHCGR in differentiating granulosa cells is mediated by intracellular levels of cAMP.

LHCGR expression is a dynamic event during the menstrual cycle and may be associated with different LH-dependent roles.[45] Desensitization is a ubiquitous phenomenon characterized by declining effector activity upon persistent agonist stimulation. The β-arrestin-dependent uncoupling and internalization is involved in the posttranscriptional regulation of the LHCGR.[46] However, the potential involvement of LHCGR phosphorylation in the process of agonist-induced internalization is not clear.

Internalized receptor forming complexes with both G proteins and β-arrestins is also assumed to induce persistent cAMP signaling, playing a crucial role in sustaining LH modulation at mid-cycle in the female.[47,48]

5. Activating mutations of the *LHCGR*

In males, activating mutations of *LHCGR* cause precocious puberty, with signs of puberty usually appearing by 3–4 years of age in boys.[21] It was also called testotoxicosis because of the early recognition of the production of toxic level of testosterone in these patients.[49] Affected boys present with penile growth, bilateral enlargement of testis, and pubic hair development. Hormone test indicated increased serum testosterone and decreased serum LH and FSH levels.[49] Excessive testosterone also induces muscular body build, accelerated growth, and advanced bone age. Histologic examination of testicular biopsy revealed nodular Leydig cell hyperplasia and seminiferous tubular development. Testicular seminoma and Leydig cell adenomas were also reported, indicating LHCGR activation may be a predisposing factor in the development of testicular tumors.[50,51] However, long-term follow-up of five cases of familial male-limited precocious puberty (FMPP) indicated testicular impairment may be present in young adulthood.[52] Interestingly, females with activating *LHCGR* mutations have no reproductive manifestations.[53] Hence, this type of "familial testotoxicosis" is inherited in an autosomal dominant, male-limited pattern.

Shenker et al.[54] identified mutation Asp578Gly in the *LHCGR* gene in affected individuals from eight different families. In vitro experiment shows LH- or hCG-independent production of cAMP or steroid hormone in cells transfected with mutant LHCGR.[50] Usually, activating mutations are located in exon 11, encoding the region responsible for signal transduction. As shown in Fig. 1, most of the activating mutations are located in the third intracellular loop (ICL) that is involved in G-protein interaction and activation. Asp578 is the "hotspot" mutated amino acid among these mutations, with over 55% of all activating mutations affecting this residue, which can be mutated to Tyr, His, Gly, and Glu.[54–61] Notably, Asp578His mutation is only found in patients with Leydig cell adenoma,[50] which is the most prevalent hormone-producing tumors of the testis and accounts for 1%–3% of all testicular tumors.[62,63] In the absence of LH or hCG, this mutant causes a strong activation of cAMP in vitro. In addition, inositol phosphate (IP) signaling pathway is also activated. Phospholipid phosphatidyl inositol

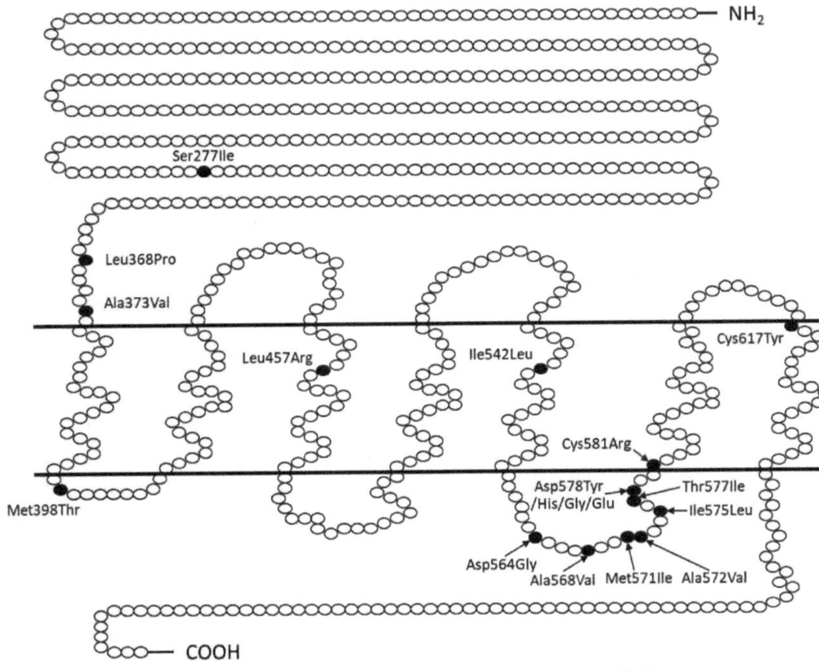

Fig. 1 Currently known activating mutations in the human *LHCGR* gene.

bisphosphate (PIP2) is broken down into IP3 and diacylglycerol (DAG), which are important for protein kinase C (PKC) activation.[57]

In 2015, Özcabı et al. reported a patient presented with peripheral precocious puberty, who has no family history. A novel *LHCGR* mutation (c.830G > T) located in exon 9 was identified. Genetic analysis of his parents was normal. Hence this mutation was probably a de novo pathogenic variant.[64] Furthermore, somatic activating mutation of *LHCGR* has also been found in tumor tissue of patients with Leydig cell adenoma. Liu et al.[56] reported the presence of an activating mutation Asp578His in the *LHCGR* gene in Leydig cell adenoma tissue of three patients. This mutation was not present in adjacent normal tissue or blood cells.

Production of cAMP in cells transfected with activating LHCGR mutant is no more than cells transfected with wild-type LHCGR when incubated with a saturating concentration of hCG, indicating that these activating mutations stabilize the LHCGR in states of intermediate activation.[21] By studying the naturally occurring and laboratory-generated activating mutations, an increase in the solvent accessibility of the cytosolic ends of TM3 and TM6 is shown to be the common structural change in activating mutations of LHCGR.[65] Amino acids affected by the disease-causing mutations of the LHCGR are in the interior of the helix bundle pointing toward the inside of the pocket formed by the seven TM helices. Moreover, LHCGR-activating mutations disrupt the salt bridge at the cytoplasmic ends of TM3 and TM6, including disruption of existing interhelical interactions and/or the introduction of new interhelical interactions.[21] In addition, electrostatic and steric interactions have also been shown to be important in receptor activation. However, the specific steps leading to LHCGR activation remain unclear.

Several drugs such as medroxyprogesterone, ketoconazole, and cyproterone have been used in the FMPP patients to reduce the synthesis and action of androgen.[64] Combination of aromatase inhibitors and antiandrogen agents showed decreased advancement of skeletal maturation, secondary sexual characteristics, and aggressiveness in the short-term period.[66] Long-term treatment with combination therapy indicated it is well tolerated, slowed bone age advancement, and prevented progression of virilization.[67] Multicenter Phase II clinical trial showed bicalutamide plus anastrozole treatment results in slower growth rate.[68]

6. Inactivating mutations of the *LHCGR*

6.1 Leydig cell hypoplasia

The human LHCGR plays a fundamental role in male and female reproductive physiology. In male embryogenesis, both hCG and LH promote the differentiation of external genitalia. After puberty, LH induces secondary sexual features. Inactivating mutations of the *LHCGR* gene in 46,XY male cause Leydig cell hypoplasia (LCH), which is a rare form of disorder of sex development characterized by a phenotype that ranges from complete 46, XY male pseudohermaphroditism to normal male sex differentiation with hypospadias or micropenis.[69,70]

LCH was first described in 1976, when genetic sequencing technique was unavailable. Berthezène et al.[71] described a 35-year-old 46,XY woman presented with completely female external genitalia. Physical examination showed she had testes, epididymis, and vas deference. However, Mullerian duct structure was absent. She showed elevated serum LH and decreased testosterone level. Furthermore, serum testosterone was unresponsive to hCG stimulation. Histopathology analysis of the removed gonads found seminiferous tubules containing normal Sertoli cells and immature Leydig cells, which led to the disorder termed LCH.[72–74] Then LCH was speculated to be caused by inactivating mutation of *LHCGR* gene or the signaling pathway downstream of the receptor. In 1995, two 46,XY siblings born from consanguineous parents were reported to present with female external genitalia, primary amenorrhea, and lack of breast development during puberty. Genetic analysis revealed the proband carried an Ala593Pro homozygous missense mutation in TM6 of *LHCGR*.[69] Her sister was also found to harbor the same mutation, even though she presented with a milder phenotype.[75]

Most inactivating mutations are missense mutations that introduced amino acid substitutions in *LHCGR*.[76] Moreover, nonsense mutations, insertions, and deletions throughout the whole gene have also been reported.[8,69,70,77] Type I LCH is the severe form of LCH caused by *LHCGR* mutations which cause the mutant LHCGR completely unresponsive to hCG and LH, resulting in female external phenotype with a blind-ending vagina, lack of breast development, and primary amenorrhea in 46,XY subjects. Type II LCH is a milder form of LCH caused by *LHCGR* mutations that retain partial receptor activity. These 46,XY individuals exhibit a range of incomplete virilization of external genitalia, such as micropenis and/or hypospadias.[21]

LCH is a rare form of recessive inheritance disease that affected patients with homozygous and compound heterozygous *LHCGR* mutations. A total of 47 inactivating mutations were identified, including missense mutations, nonsense mutations, insertions, and deletions (Fig. 2). Single base substitutions leading to either missense mutations or nonsense mutations have been identified from ECD, TM, and cytoplasmic loops. A 33-bp inframe insertion has been found in the maternal *LHCGR* allele in a kindred with two 46, XY male pseudohermaphrodites.[78] A homozygous deletion of approximately 5 kb encompassing exon 10 of the *LHCGR* gene was identified from a patient with partial LCH, manifesting with retarded pubertal development, small testicles, and delayed bone maturation at 18 years old. This patient displays nearly normal male phenotype but lacks pubertal development, with elevated LH level and decreased testosterone level. An hCG stimulation test induces testosterone biosynthesis and secretion within the normal range, suggesting a possible dual mechanism of hormone binding and signal transduction for hCG and LH on an *LHCGR* lacking exon 10.[79]

The ECD of the LHCGR is responsible for high-affinity binding of the hormone, which initiates signal transduction. Four mutations located in the LRRs of extracellular ligand-binding domain of the receptor, namely, Ile114Phe, Cys131Arg, Val144Phe, and Phe194Val, were identified from LCH patients. Among them, Val144Phe and Phe194Val are observed

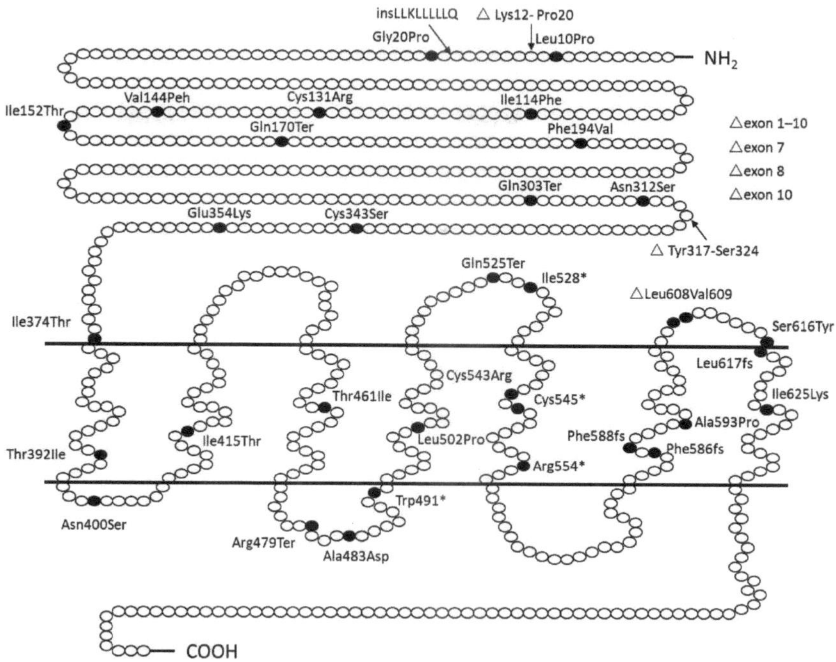

Fig. 2 Currently known inactivating mutations in the human *LHCGR* gene.

to interfere with membrane transportation of the receptor,[80,81] whereas Ile114Phe in the LRR3 results in reduced ligand binding and signal transduction of the receptor, with less effect on membrane insertion of the receptor.[82] Ile152Thr was observed to cause LCH, with the 46,XY girl showing unresponsive T to hCG, histologically immature seminiferous tubules. and lack of Leydig cells in the gonads (Fig. 3A and B). This mutation, located in the LRR5, was demonstrated to abolish the binding of the hormone and signal transduction. Similar to Ile114Phe, cell surface expression of the Ile152Thr mutant receptor is not disturbed. However, cells transfected with both mutants exhibit reduced hormone response, as shown by the studies with intact cells and detergent-solubilized cell lysates. Recently, a novel L10P mutation located in the signal peptide of LHCGR, combined with missense mutation S616Y, was identified from a patient with micropenis and cryptorchidism. Functional study showed that L10P variant does not completely abolish LHCGR activity but affects receptor expression.[83]

Six splicing site mutations were reported. A splicing mutation was discovered to be a C to A transition at the −3 position of the consensus sequence of the intron 6 splice acceptor site (i.e., c.537-3C>A) from a

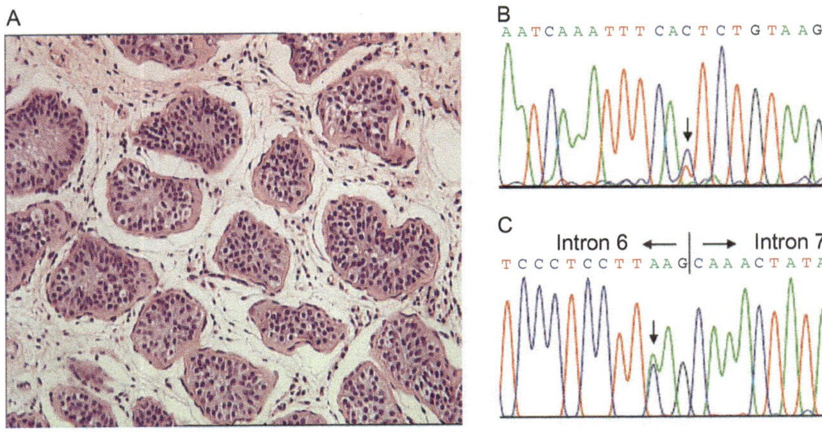

Fig. 3 (A) Histologic analysis revealed immature seminiferous tubules and arrested sperm maturation. The interstitium contains fibroblast-like cells but no mature Leydig cells. (B, C) Schematic representation of missense mutation of I152T and splice acceptor site mutation at intron 6 of *LHCGR* gene from a patient with LCH.

46,XY patient with disorder of sex development (Fig. 3C). In vitro functional splicing assay found that c.537-3C > A decreased the functionality of 3′ splice acceptor site, leading to skipping of exon 7 in the mature mRNA of the *LHCGR* gene.[84] It is interesting to note that c.557A > C, c.558G > C, and c.580A > G are located in intron 6 and can result in the aberrant splicing of exon 6A.[22,85] Mutation c.681-1G > A, located in intron 8, can lead to aberrant splicing products resulting from alternative splicing skipping or shortening of exon 9.[86] Splicing site mutation c.955-1G > A is located in intron 10. The male patient presents with delayed puberty, micropenis, and oligospermia. In vitro analysis showed that the splicing mutation causes eight amino acids deletion in the ECD (Tyr317-Ser324).[87]

The phenotype of females with inactivating *LHCGR* mutations is also heterogeneous. A homozygous deletion of Leu608 and Val609 in TM7 is demonstrated to be associated with familial testicular and ovarian resistance, due to both decreased cell surface expression and impaired effector activation by the cell surface receptor.[88] Toledo et al.[89] reported 46,XX females with *LHCGR*-inactivating mutations presented with amenorrhea due to hypergonadotropic hypogonadism but with structurally normal ovaries.

6.2 Empty follicle syndrome

Empty follicle syndrome (EFS) is defined as the failure to retrieve oocytes from mature ovarian follicles after ovulation induction for in vitro fertilization (IVF) after ovarian stimulation, despite normal follicle development.[90] Stevenson and Lashen[91] classified EFS as either "false" or "genuine" according to the demonstration of the optimum levels of β-hCG on the day of oocyte retrieval. Hence "genuine EFS" was referred as failure to retrieve oocytes from mature ovarian follicles after ovulation induction after normal follicular development and steroidogenesis in the presence of optimum β-hCG levels on the day of oocyte retrieval.

In 2011, Yariz et al.[92] reported that two affected sisters, diagnosed with genuine EFS, were identified with novel missense mutation Asn400Ser in *LHCGR* gene. From three-dimensional modeling, the highly conserved asparagine at residue 400 interacts with leucine at position 452, methionine at position 487, and tryptophan at position 491. A mutation (tryptophan) of this residue reduces receptor expression level to only 32% that of wild-type protein and hCG-saturation binding level is also impaired.[93] Recently, two Chinese patients with EFS from unrelated consanguineous families were shown to have two novel homozygous inactivating nonsense mutations in *LHCGR*: c.736C>T (Gln246X) and c.846dupT (Arg283X). Both patients had very low E2 production and uncontrolled E2 decline during follicular growth, but had acceptable follicle number and size. One patient has a sibling present with 46,XY disorder of sex development, showing inguinal testis, no uterus and ovaries.[94]

6.3 Diseases associated with polymorphisms of *LHCGR* gene

There are more than 1000 SNPs in *LHCGR* (https://www.ncbi.nlm.nih.gov/SNP/), most of which are located in the introns. Simoni et al.[95] studied the relationship between polymorphisms of *LHCGR* gene and maldescended testes. In 278 patients with maldescended testes, rs2293275 polymorphism (Ser312Asn) in exon 10 of *LHCGR* gene is significantly associated with spermatogenetic damage rather than with the maldescended testes. Ingles et al.[96] found Ser312Asn is associated with increased prostate cancer mortality in men from United States. Furthermore, this polymorphism is also related to high T stages in prostate cancer.[97] Testicular cancer may in part be caused by an imbalance of androgen/estrogen levels. Kristiansen et al.[98] found the association between Ser312Asn polymorphism and risk of developing testicular cancer. Researchers indicated two SNPs (rs4597581 and rs4953617) in the *LHCGR* gene are associated with risk of testicular germ cell cancer in Swedish and Danish patients.[99] Piersma et al. showed that the insLQ polymorphism in exon 1 of *LHCGR* is not associated with breast cancer risk. However, it is significantly associated with a shorter disease-free survival in breast cancer patients. Functional studies manifested that LHCGR-insLQ can increase receptor sensitivity and plasma membrane expression, thus increased LHCGR activity.[100]

A cryptic exon (exon 6A) was found to be derived from alternative splicing in 2.7 kb intron 6, which is only present in primates and humans.[22] We investigated the function of SNP (rs68073206) near the splicing donor site of exon 6A in azoospermia, oligoasthenozoospermia, and normozoospermia subjects. It indicated that rs68073206 shows no association with male infertility. However, this polymorphism is significantly associated with testosterone levels in normozoospermia subjects.[23]

Polycystic ovary syndrome (PCOS) is defined by the presence of anovulation, hyperandrogenism, and polycystic ovary morphology. In females, LHCGR is involved in the production of androgens, which are converted to estrogens by aromatase.[50] Therefore, the relationship between PCOS and polymorphism of *LHCGR* was hypothesized. In 2010, GWAS and subsequent replication studies identified *LHCGR* gene SNP (rs13405728) as a new susceptibility locus for PCOS in Han Chinese women.[101] After that, rs13405728 was found strongly associated with anovulation after adjustment for age and BMI.[102] In addition, Almawi et al.[103] confirmed association of *LHCGR* SNPs (rs7371084, rs4953616) with PCOS in Bahraini Arab women. Mutharasan et al.[104] also detected strong association of PCOS with *LHCGR* SNPs (rs7562215 and rs10495960) in women of European ancestry. Although *LHCGR* insLQ polymorphism has lower frequency in PCOS patients, this SNP is not significantly correlated with PCOS after Bonferroni's correction.[105] Liu et al.[106] found a novel *LHCGR* SNP near exon 7 (NT27762960). However, this SNP is not significantly associated with PCOS.

7. Potential therapeutic strategies

During the past two decades, some small-molecule orthosteric ligands (i.e., pharmacological chaperones) were shown to interact with intracellularly retained mutant GPCRs to aid their folding and transport to the plasma membrane (reviewed in refs. 107,108). Small molecules of permeable vasopressin analogs that act as pharmacological chaperones increase the cell surface expression of mutant V2 vasopressin receptors.[109] Moreover, nonpeptide antagonist SR49059 improves symptoms (decreased urine volume and water intake) in patients with nephrogenic diabetes insipidus harboring transport-defective V2 vasopressin receptor mutations.[110] In 2011, Newton et al.[111] demonstrated that thienopyrimidine, Org-42599, rescues plasma membrane expression and thereby signaling of two intracellularly retained mutant LHCGRs (Asp593Pro and Ser616Tyr), which were identified in LCH patients. Although Org-42599 can activate these receptors, it cannot displace ^{125}I-labeled LH binding to the wild-type receptor, indicating that it acts in an allosteric manner. It is not clear whether this would be the case for all ER-retained mutant LHCGRs. Allosteric agonists make it possible for treatment of infertile patients with *LHCGR* mutations that cause intracellular retention of the misfolded receptors.

References

1. Ascoli M, Fanelli F, Segaloff DL. The lutropin/choriogonadotropin receptor, a 2002 perspective. *Endocr Rev*. 2002;23:141–174.
2. Liu JH, Yen SSC. Induction of midcycle gonadotropin surge by ovarian steroids in women: a critical evaluation. *J Clin Endocrinol Metab*. 1983;57:797–802.
3. Vande Wiele RL, Bogumil J, Dyrenfurth I, et al. Mechanisms regulating the menstrual cycle in women. *Recent Prog Horm Res*. 1970;26:63–103.
4. Duncan WC, McNeilly AS, Fraser HM, Illingworth PJ. Luteinizing hormone receptor in the human corpus luteum: lack of down-regulation during maternal recognition of pregnancy. *Hum Reprod*. 1996;11:2291–2297.
5. Hutchison JS, Nelson PB, Zeleznik AJ. Effects of different gonadotropin pulse frequencies on corpus luteum function during the menstrual cycle of rhesus monkeys. *Endocrinology*. 1986;119:1964–1971.
6. Shi QJ, Lei ZM, Rao CV, Lin J. Novel role of human chorionic gonadotropin in differentiation of human cytotrophoblasts. *Endocrinology*. 1993;132:1387–1395.
7. Tao YX, Lei ZM, Hofmann GE, Rao CV. Human intermediate trophoblasts express chorionic gonadotropin/luteinizing hormone receptor gene. *Biol Reprod*. 1995;53: 899–904.
8. Themmen APN, Huhtaniemi IT. Mutations of gonadotropins and gonadotropin receptors: elucidating the physiology and pathophysiology of pituitary-gonadal function. *Endocr Rev*. 2000;21:551–583.
9. Yamada G, Suzuki K, Haraguchi R, et al. Molecular genetic cascades for external genitalia formation: an emerging organogenesis program. *Dev Dyn*. 2006;235:1738–1752.
10. Pierce JG, Parsons TF. Glycoprotein hormones: structure and function. *Annu Rev Biochem*. 1981;50:465–495.
11. Bousfield GR, Butnev VY, Butnev VY. Identification of twelve O-glycosylation sites in equine chorionic gonadotropin beta and equine luteinizing hormone ss by solid-phase Edman degradation. *Biol Reprod*. 2001;64:136–147.
12. Stewart F, Allen WR. Biological functions and receptor binding activities of equine chorionic gonadotrophins. *J Reprod Fertil*. 1981;62:527–536.
13. Simula AP, Amato F, Faast R, Lopata A, Berka J, Norman RJ. Luteinizing hormone/ chorionic gonadotropin bioactivity in the common marmoset (*Callithrix jacchus*) is due to a chorionic gonadotropin molecule with a structure intermediate between human chorionic gonadotropin and human luteinizing hormone. *Biol Reprod*. 1995;53: 380–389.
14. Rahman NA, Rao CV. Recent progress in luteinizing hormone/human chorionic gonadotrophin hormone research. *Mol Hum Reprod*. 2009;15:703–711.

15. Rao CV. Differential properties of human chorionic gonadotrophin and human luteinizing hormone binding to plasma membranes of bovine corpora lutea. *Acta Endocrinol.* 1979;90:696–710.
16. Cole LA. New discoveries on the biology and detection of human chorionic gonadotropin. *Reprod Biol Endocrinol.* 2009;7:8.
17. Zhang FP, Poutanen M, Wilbertz J, Huhtaniemi I. Normal prenatal but arrested postnatal sexual development of luteinizing hormone receptor knockout (LuRKO) mice. *Mol Endocrinol.* 2001;15:172–183.
18. Lei ZM, Mishra S, Zou W, et al. Targeted disruption of luteinizing hormone/human chorionic gonadotropin receptor gene. *Mol Endocrinol.* 2001;15:184–200.
19. Pakarainen T, Zhang FP, Nurmi L, Poutanen M, Huhtaniemi I. Knockout of luteinizing hormone receptor abolishes the effects of follicle-stimulating hormone on preovulatory maturation and ovulation of mouse graafian follicles. *Mol Endocrinol.* 2005;19:2591–2602.
20. Segaloff DL, Ascoli M. The lutropin/choriogonadotropin receptor 4 years later. *Endocr Rev.* 1993;14:324–347.
21. Segaloff DL. Diseases associated with mutations of the human lutropin receptor. *Prog Mol Biol Transl Sci.* 2009;89:97–114.
22. Kossack N, Simoni M, Richter-Unruh A, Themmen AP, Gromoll J. Mutations in a novel, cryptic exon of the luteinizing hormone/chorionic gonadotropin receptor gene cause male pseudohermaphroditism. *PLoS Med.* 2008;5, e88.
23. Liu W, Han B, Zhu W, et al. Polymorphism in the alternative donor site of the cryptic exon of LHCGR: functional consequences and associations with testosterone level. *Sci Rep.* 2017;7, 45699.
24. Fan QR, Hendrickson WA. Structure of human follicle-stimulating hormone in complex with its receptor. *Nature.* 2005;433:269–277.
25. Vischer HF, Granneman JC, Noordam MJ, Mosselman S, Bogerd J. Ligand selectivity of gonadotropin receptors. Role of the beta-strands of extracellular leucine-rich repeats 3 and 6 of the human luteinizing hormone receptor. *J Biol Chem.* 2003;278: 15505–15513.
26. Angelova K, de Jonge H, Granneman JC, Puett D, Bogerd J. Functional differences of invariant and highly conserved residues in the extracellular domain of the glycoprotein hormone receptors. *J Biol Chem.* 2010;285:34813–34827.
27. Bhowmick N, Huang J, Puett D, Isaacs NW, Lapthorn AJ. Determination of residues important in hormone binding to the extracellular domain of the luteinizing hormone/chorionic gonadotropin receptor by site-directed mutagenesis and modeling. *Mol Endocrinol.* 1996;10:1147–1159.
28. Bhowmick N, Narayan P, Puett D. Identification of ionizable amino acid residues on the extracellular domain of the lutropin receptor involved in ligand binding. *Endocrinology.* 1999;140:4558–4563.
29. Tao YX, Johnson NB, Segaloff DL. Constitutive and agonist-dependent self-association of the cell surface human lutropin receptor. *J Biol Chem.* 2004;279: 5904–5914.
30. Bradbury FA, Kawate N, Foster CM, Menon KM. Post-translational processing in the Golgi plays a critical role in the trafficking of the luteinizing hormone/human chorionic gonadotropin receptor to the cell surface. *J Biol Chem.* 1997;272:5921–5926.
31. Jiang X, Liu H, Chen X, et al. Structure of follicle-stimulating hormone in complex with the entire ectodomain of its receptor. *Proc Natl Acad Sci USA.* 2012;109: 12491–12496.
32. Bonomi M, Busnelli M, Persani L, Vassart G, Costagliola S. Structural differences in the hinge region of the glycoprotein hormone receptors: evidence from the sulfated tyrosine residues. *Mol Endocrinol.* 2006;20:3351–3363.

33. Bruysters M, Verhoef-Post M, Themmen AP. Asp330 and Tyr331 in the C-terminal cysteine-rich region of the luteinizing hormone receptor are key residues in hormone-induced receptor activation. *J Biol Chem*. 2008;283:25821–25828.
34. Troppmann B, Kleinau G, Krause G, Gromoll J. Structural and functional plasticity of the luteinizing hormone/choriogonadotrophin receptor. *Hum Reprod Update*. 2013;19:583–602.
35. Casarini L, Santi D, Brigante G, Simoni M. Two hormones for one receptor: evolution, biochemistry, actions and pathophysiology of LH and hCG. *Endocr Rev*. 2018;39: 549–592.
36. Gudermann T, Nichols C, Levy FO, Birnbaumer M, Birnbaumer L. Ca^{2+} mobilization by the LH receptor expressed in *Xenopus* oocytes independent of 3′,5′-cyclic adenosine monophosphate formation: evidence for parallel activation of two signaling pathways. *Mol Endocrinol*. 1992;6:272–278.
37. Zhu X, Gilbert S, Birnbaumer M, Birnbaumer L. Dual signaling potential is common among Gs-coupled receptors and dependent on receptor density. *Mol Pharmacol*. 1994;46:460–469.
38. Ben-Ami I, Armon L, Freimann S, Strassburger D, Ron-El R, Amsterdam A. EGF-like growth factors as LH mediators in the human corpus luteum. *Hum Reprod*. 2009;24:176–184.
39. Palaniappan M, Menon KM. Human chorionic gonadotropin stimulates theca-interstitial cell proliferation and cell cycle regulatory proteins by a cAMP-dependent activation of AKT/mTORC1 signaling pathway. *Mol Endocrinol*. 2010;24:1782–1793.
40. Choi J, Smitz J. Luteinizing hormone and human chorionic gonadotropin: origins of difference. *Mol Cell Endocrinol*. 2014;383:203–213.
41. Casarini L, Lispi M, Longobardi S, et al. LH and hCG action on the same receptor results in quantitatively and qualitatively different intracellular signalling. *PLoS One*. 2012;7, e46682.
42. Gupta C, Chapekar T, Chhabra Y, Singh P, Sinha S, Luthra K. Differential response to sustained stimulation by hCG & LH on goat ovarian granulosa cells. *Indian J Med Res*. 2012;135:331–340.
43. Tsai-Morris CH, Geng Y, Buczko E, Dufau ML. Characterization of diverse functional elements in the upstream Sp1 domain of the rat luteinizing hormone receptor gene promoter. *J Biol Chem*. 1995;270:7487–7494.
44. Tsai-Morris CH, Geng Y, Xie XZ, Buczko E, Dufau ML. Transcriptional protein binding domains governing basal expression of the rat luteinizing hormone receptor gene. *J Biol Chem*. 1994;269:15868–15875.
45. Jeppesen JV, Kristensen SG, Nielsen ME, et al. LH-receptor gene expression in human granulosa and cumulus cells from antral and preovulatory follicles. *J Clin Endocrinol Metab*. 2012;97:E1524–E1531.
46. Hunzicker-Dunn M, Barisas G, Song J, Roess DA. Membrane organization of luteinizing hormone receptors differs between actively signaling and desensitized receptors. *J Biol Chem*. 2003;278:42744–42749.
47. Thomsen ARB, Plouffe B, Cahill 3rd TJ, et al. GPCR-G protein-β-arrestin super-complex mediates sustained G protein signaling. *Cell*. 2016;166:907–919.
48. Lyga S, Volpe S, Werthmann RC, et al. Persistent cAMP signaling by internalized LH receptors in ovarian follicles. *Endocrinology*. 2016;157:1613–1621.
49. Rosenthal SM, Grumbach MM, Kaplan SL. Gonadotropin independent familial sexual precocity with premature Leydig and germinal cell maturation (familial testotoxicosis): effects of a potent luteinizing hormone-releasing factor agonist and medroxyprogesterone acetate therapy in four cases. *J Clin Endocrinol Metab*. 1983;57:571–579.
50. Piersma D, Verhoef-Post M, Berns EM, Themmen AP. LH receptor gene mutations and polymorphisms: an overview. *Mol Cell Endocrinol*. 2007;260–262:282–286.

51. Martin MM, Wu SM, Martin AL, Rennert OM, Chan WY. Testicular seminoma in a patient with a constitutively activating mutation of the luteinizing hormone/chorionic gonadotropin receptor. *Eur J Endocrinol.* 1998;139:101–106.
52. Bertelloni S, Baroncelli GI, Lala R, et al. Long-term outcome of male-limited gonadotropin-independent precocious puberty. *Horm Res.* 1997;48:235–239.
53. Rosenthal IM, Refetoff S, Rich B, et al. Response to challenge with gonadotropin-releasing hormone agonist in a mother and her two sons with a constitutively activating mutation of the luteinizing hormone receptor—a clinical research center study. *J Clin Endocrinol Metab.* 1996;81:3802–3806.
54. Shenker A, Laue L, Kosugi S, Merendino Jr JJ, Minegishi T, Cutler Jr GB. A constitutively activating mutation of the luteinizing hormone receptor in familial male precocious puberty. *Nature.* 1993;365:652–654.
55. Muller J, Gondos B, Kosugi S, Mori T, Shenker A. Severe testotoxicosis phenotype associated with Asp578–>Tyr mutation of the lutrophin/choriogonadotrophin receptor gene. *J Med Genet.* 1998;35:340–341.
56. Liu G, Duranteau L, Carel JC, Monroe J, Doyle DA, Shenker A. Leydig-cell tumors caused by an activating mutation of the gene encoding the luteinizing hormone receptor. *N Engl J Med.* 1999;341:1731–1736.
57. Richter-Unruh A, Wessels HT, Menken U, et al. Male LH-independent sexual precocity in a 3.5-year-old boy caused by a somatic activating mutation of the LH receptor in a Leydig cell tumor. *J Clin Endocrinol Metab.* 2002;87:1052–1056.
58. Wang M, Li M, Liu YS, Lei SM, Xiao YF. Familial male-limited precocious puberty due to Asp578His mutations in the LHCGR gene: clinical characteristics and gene analysis in an infant. *Zhongguo Dang Dai Er Ke Za Zhi.* 2017;19:1159–1164.
59. Siviero-Miachon AA, Kizys MM, Ribeiro MM, Garcia FE, Spinola-Castro AM, Dias da Silva MR. Cosegregation of a novel mutation in the sixth transmembrane segment of the luteinizing/choriogonadotropin hormone receptor with two Brazilian siblings with severe testotoxicosis. *Endocr Res.* 2017;42:117–124.
60. Yano K, Hidaka A, Saji M, et al. A sporadic case of male-limited precocious puberty has the same constitutively activating point mutation in luteinizing hormone/choriogonadotropin receptor gene as familial cases. *J Clin Endocrinol Metab.* 1994;79:1818–1823.
61. Wu SM, Leschek EW, Brain C, Chan WY. A novel luteinizing hormone receptor mutation in a patient with familial male-limited precocious puberty: effect of the size of a critical amino acid on receptor activity. *Mol Genet Metab.* 1999;66:68–73.
62. Kim I, Young RH, Scully RE. Leydig cell tumors of the testis: a clinicopathological analysis of 40 cases and review of the literature. *Am J Surg Pathol.* 1985;9:177–192.
63. Scully RE. Testicular tumors with endocrine manifestations. In: DeGroot LJ, ed. *Endocrinology.* 3rd ed. Philadelphia: W.B. Saunders; 1995:2442–2448.
64. Özcabı B, TahmiscioğluBucak F, Ceylaner S, et al. Testotoxicosis: report of two cases, one with a novel mutation in LHCGR gene. *J Clin Res Pediatr Endocrinol.* 2015;7:242–248.
65. Fanelli F, Verhoef-Post M, Timmerman M, Zeilemaker A, Martens JW, Themmen AP. Insight into mutation-induced activation of the luteinizing hormone receptor: molecular simulations predict the functional behavior of engineered mutants at M398. *Mol Endocrinol.* 2004;18:1499–1508.
66. Mitre N, Lteif A. Treatment of familial male-limited precocious puberty (testotoxicosis) with anastrozole and bicalutamide in a boy with a novel mutation in the luteinizing hormone receptor. *J Pediatr Endocrinol Metab.* 2009;22:1163–1167.
67. Lenz AM, Shulman D, Eugster EA, et al. Bicalutamide and third generation aromatase inhibitors in testotoxicosis. *Pediatrics.* 2010;126:728–733.

68. Reiter EO, Mauras N, McCormick K, et al. Bicalutamide plus anastrozole for the treatment of gonadotropin-independent precocious puberty in boys with testotoxicosis: a phase II, open-label pilot study (BATT). *J Pediatr Endocrinol Metab.* 2010;23:999–1009.
69. Kremer H, Kraaij R, Toledo SP, et al. Male pseudohermaphroditism due to a homozygous missense mutation of the luteinizing hormone receptor gene. *Nat Genet.* 1995;9:160–164.
70. Latronico AC, Anasti J, Arnhold IJ, et al. Brief report: testicular and ovarian resistance to luteinizing hormone caused by inactivating mutations of the luteinizing hormone-receptor gene. *N Engl J Med.* 1996;334:507–512.
71. Berthezène F, Forest MG, Grimaud JA, Claustrat B, Mornex R. Leydig-cell agenesis: a cause of male pseudohermaphroditism. *N Engl J Med.* 1976;295:969–972.
72. Brown DM, Markland C, Dehner LP. Leydig cell hypoplasia: a cause of male pseudohermaphroditism. *J Clin Endocrinol Metab.* 1978;46:1–7.
73. Wu RH, Rosenfeld R, Fukushima D. Hypogonadism and Leydig cell hypoplasia unresponsive to human luteinizing hormone (hLH). *Am J Med Sci.* 1984;287:23–25.
74. Eil C, Austin RM, Sesterhenn I, Dunn JF, Cutler Jr GB, Johnsonbaugh RE. Leydig-cell hypoplasia causing male pseudohermaphroditism: diagnosis 13 years after prepubertal castration. *J Clin Endocrinol Metab.* 1984;58:441–448.
75. Rousseau-Merck MF, Misrahi M, Atger M, Loosfelt H, Milgrom E, Berger R. Localization of the human luteinizing hormone/choriogonadotropin receptor gene (LHCGR) to chromosome 2p21. *Cytogenet Cell Genet.* 1990;54:77–79.
76. Huhtaniemi I, Alevizaki M. Gonadotrophin resistance. *Best Pract Res Clin Endocrinol Metab.* 2006;20:561–576.
77. Laue LL, Wu SM, Kudo M, et al. Compound heterozygous mutations of the luteinizing hormone receptor gene in Leydig cell hypoplasia. *Mol Endocrinol.* 1996;10: 987–997.
78. Wu SM, Hallermeier KM, Laue L, et al. Inactivation of the luteinizing hormone/chorionic gonadotropin receptor by an insertional mutation in Leydig cell hypoplasia. *Mol Endocrinol.* 1998;12:1651–1660.
79. Gromoll J, Eiholzer U, Nieschlag E, Simoni M. Male hypogonadism caused by homozygous deletion of exon 10 of the luteinizing hormone (LH) receptor: differential action of human chorionic gonadotropin and LH. *J Clin Endocrinol Metab.* 2000;85: 2281–2286.
80. Richter-Unruh A, Verhoef-Post M, Malak S, Homoki J, Hauffa BP, Themmen AP. Leydig cell hypoplasia: absent luteinizing hormone receptor cell surface expression caused by a novel homozygous mutation in the extracellular domain. *J Clin Endocrinol Metab.* 2004;89:5161–5167.
81. Gromoll J, Lahrmann L, Godmann M, et al. Genomic checkpoints for exon 10 usage in the luteinizing hormone receptor type 1 and type 2. *Mol Endocrinol.* 2007;21:1984–1996.
82. Leung MY, Steinbach PJ, Bear D, et al. Biological effect of a novel mutation in the third leucine-rich repeat of human luteinizing hormone receptor. *Mol Endocrinol.* 2006;20:2493–2503.
83. Vezzoli V, Duminuco P, Vottero A, et al. A new variant in signal peptide of the human luteinizing hormone receptor (LHCGR) affects receptor biogenesis causing Leydig cell hypoplasia. *Hum Mol Genet.* 2015;24:6003–6012.
84. Qiao J, Han B, Liu BL, et al. A splice site mutation combined with a novel missense mutation of LHCGR cause male pseudohermaphroditism. *Hum Mutat.* 2009;30: E855–E865.
85. Kossack N, Troppmann B, Richter-Unruh A, Kleinau G, Gromoll J. Aberrant transcription of the LHCGR gene caused by a mutation in exon 6A leads to Leydig cell hypoplasia type II. *Mol Cell Endocrinol.* 2013;366:59–67.

86. Xu Y, Chen Y, Li N, et al. Novel compound heterozygous variants in the LHCGR gene identified in a subject with Leydig cell hypoplasia type 1. *J Pediatr Endocrinol Metab.* 2018;31:239–245.
87. Bruysters M, Christin-Maitre S, Verhoef-Post M, et al. A new LH receptor splice mutation responsible for male hypogonadism with subnormal sperm production in the propositus, and infertility with regular cycles in an affected sister. *Hum Reprod.* 2008;23:1917–1923.
88. Latronico AC, Chai Y, Arnhold IJ, Liu X, Mendonca BB, Segaloff DL. A homozygous microdeletion in helix 7 of the luteinizing hormone receptor associated with familial testicular and ovarian resistance is due to both decreased cell surface expression and impaired effector activation by the cell surface receptor. *Mol Endocrinol.* 1998; 12:442–450.
89. Toledo SP, Brunner HG, Kraaij R, et al. An inactivating mutation of the luteinizing hormone receptor causes amenorrhea in a 46,XX female. *J Clin Endocrinol Metab.* 1996;81:3850–3854.
90. Revelli A, Carosso A, Grassi G, Gennarelli G, Canosa S, Benedetto C. Empty follicle syndrome revisited: definition, incidence, aetiology, early diagnosis and treatment. *Reprod Biomed Online.* 2017;35:132–138.
91. Stevenson TL, Lashen H. Empty follicle syndrome: the reality of a controversial syndrome, a systematic review. *Fertil Steril.* 2008;90:691–698.
92. Yariz KO, Walsh T, Uzak A, et al. Inherited mutation of the luteinizing hormone/ choriogonadotropin receptor (LHCGR) in empty follicle syndrome. *Fertil Steril.* 2011;96:e125–e130.
93. Angelova K, Felline A, Lee M, Patel M, Puett D, Fanelli F. Conserved amino acids participate in the structure networks deputed to intramolecular communication in the lutropin receptor. *Cell Mol Life Sci.* 2011;68:1227–1239.
94. Chen C, Xu X, Kong L, et al. Novel homozygous nonsense mutations in LHCGR lead to empty follicle syndrome and 46,XY disorder of sex development. *Hum Reprod.* 2018;33:1364–1369.
95. Simoni M, Tüttelmann F, Michel C, Böckenfeld Y, Nieschlag E, Gromoll J. Polymorphisms of the luteinizing hormone/chorionic gonadotropin receptor gene: association with maldescended testes and male infertility. *Pharmacogenet Genomics.* 2008;18: 193–200.
96. Ingles SA, Liu SV, Pinski J. LHRH and LHR genotypes and prostate cancer incidence and survival. *Int J Mol Epidemiol Genet.* 2013;4:228–234.
97. Robles-Fernandez I, Martinez-Gonzalez LJ, Pascual-Geler M, et al. Association between polymorphisms in sex hormones synthesis and metabolism and prostate cancer aggressiveness. *PLoS One.* 2017;12, e0185447.
98. Kristiansen W, Aschim EL, Andersen JM, Witczak O, Fosså SD, Haugen TB. Variations in testosterone pathway genes and susceptibility to testicular cancer in Norwegian men. *Int J Androl.* 2012;35:819–827.
99. Brokken LJ, Lundberg-Giwercman Y, Rajpert De-Meyts E, et al. Association of polymorphisms in genes encoding hormone receptors ESR1, ESR2 and LHCGR with the risk and clinical features of testicular germ cell cancer. *Mol Cell Endocrinol.* 2012;351:279–285.
100. Piersma D, Berns EM, Verhoef-Post M, et al. A common polymorphism renders the luteinizing hormone receptor protein more active by improving signal peptide function and predicts adverse outcome in breast cancer patients. *J Clin Endocrinol Metab.* 2006;91:1470–1476.
101. Chen ZJ, Zhao H, He L, et al. Genome-wide association study identifies susceptibility loci for polycystic ovary syndrome on chromosome 2p16.3, 2p21 and 9q33.3. *Nat Genet.* 2011;43:55–59.

102. Cui L, Li G, Zhong W, et al. Polycystic ovary syndrome susceptibility single nucleotide polymorphisms in women with a single PCOS clinical feature. *Hum Reprod.* 2015; 30:732–736.
103. Almawi WY, Hubail B, Arekat DZ, et al. Luteinizing hormone/choriogonadotropin receptor and follicle stimulating hormone receptor gene variants in polycystic ovary syndrome. *J Assist Reprod Genet.* 2015;32:607–614.
104. Mutharasan P, Galdones E, Peñalver Bernabé B, et al. Evidence for chromosome 2p16.3 polycystic ovary syndrome susceptibility locus in affected women of European ancestry. *J Clin Endocrinol Metab.* 2013;98:E185–E190.
105. Valkenburg O, Uitterlinden AG, Piersma D, et al. Genetic polymorphisms of GnRH and gonadotrophic hormone receptors affect the phenotype of polycystic ovary syndrome. *Hum Reprod.* 2009;24:2014–2022.
106. Liu N, Ma Y, Wang S, et al. Association of the genetic variants of luteinizing hormone, luteinizing hormone receptor and polycystic ovary syndrome. *Reprod Biol Endocrinol.* 2012;10:36.
107. Tao YX, Conn PM. Chaperoning G protein-coupled receptors: from cell biology to therapeutics. *Endocr Rev.* 2014;35:602–647.
108. Tao YX, Conn PM. Pharmacoperones as novel therapeutics for diverse protein conformational diseases. *Physiol Rev.* 2018;98:697–725.
109. Morello JP, Salahpour A, Laperrière A, et al. Pharmacological chaperones rescue cell-surface expression and function of misfolded V2 vasopressin receptor mutants. *J Clin Invest.* 2000;105:887–895.
110. Bernier V, Morello JP, Zarruk A, et al. Pharmacologic chaperones as a potential treatment for X-linked nephrogenic diabetes insipidus. *J Am Soc Nephrol.* 2006;17:232–243.
111. Newton CL, Whay AM, McArdle CA, et al. Rescue of expression and signaling of human luteinizing hormone G protein-coupled receptor mutants with an allosterically binding small-molecule agonist. *Proc Natl Acad Sci USA.* 2011;108:7172–7176.

CHAPTER FIVE

Physiology and pathophysiology of the β_3-adrenergic receptor

Li-Kun Yang, Ya-Xiong Tao*
Department of Anatomy, Physiology and Pharmacology, College of Veterinary Medicine, Auburn University, Auburn, AL, United States
*Corresponding author: e-mail address: taoyaxi@auburn.edu

Contents

Abstract

The β_3-adrenergic receptor (β_3-AR) is an important regulator of various physiological functions, such as thermogenesis in brown adipose tissue, lipolysis in white adipose tissue, negative inotropic effect in cardiomyocyte, and relaxation in blood vessel. The activation of β_3-AR by its agonists is shown to have metabolic (antiobesity and anti-diabetic) and cardiovascular effects in animal models, highlighting β_3-AR as a potential therapeutic target in the treatment of several human diseases. Moreover, a substantial number of studies performed on different populations have identified some β_3-AR polymorphic variants associated with obesity, diabetes, cardiovascular diseases, and other disorders. The clinical phenotypes and functional characteristics of these variants provide insights into potential pathophysiological roles of β_3-AR in the development of these diseases.

Progress in Molecular Biology and Translational Science, Volume 161
ISSN 1877-1173
https://doi.org/10.1016/bs.pmbts.2018.09.003

Abbreviations

AC adenylyl cyclase
ADRB3 adrenoceptor β3
ApoE4 ε4 allele of the Apolipoprotein E gene
AR adrenergic receptor
BAT brown adipose tissue
BMI body mass index
cAMP 3′,5′-cyclic adenosine monophosphate
cGMP 3′,5′-cyclic guanosine monophosphate
ECL extracellular loop
eNOS endothelial nitric oxide synthase
ERK1/2 extracellular signal-regulated kinase 1/2
Gi inhibitory G protein
GPCR G protein-coupled receptor
Gs stimulatory G-protein
ICL intracellular loop
NO nitric oxide
PI3K phosphatidyl inositol 3-kinase
PKA protein kinase A
SNP single nucleotide polymorphism
T2DM type 2 diabetes mellitus
TMD transmembrane domain
UCP-1 uncoupling protein 1
UTR untranslated region
WAT white adipose tissue

1. The β-Adrenergic receptor system

Adrenergic receptors (ARs), members of family A G protein-coupled receptors (GPCRs), are well known as central mediators in the sympathetic nervous system. By binding to catecholamines, ARs exert their effects on regulating the functions of organs and tissues to maintain homeostasis and generating fight-or-flight responses (e.g., dilating the pupil, increasing heart rate, mobilizing energy, and diverting blood flow from nonessential organs to skeletal muscle). So far, a total of nine subtypes of ARs have been identified based on their structural and functional differences, which are divided into three subfamilies: α_1-ARs (α_{1A}-AR, α_{1B}-AR, and α_{1D}-AR), α_2-ARs (α_{2A}-AR, α_{2B}-AR, and α_{2C}-AR), and β-ARs (β_1-AR, β_2-AR, and β_3-AR).

1.1 The endogenous ligands of β-ARs

The endogenous ligands for β-ARs are norepinephrine and epinephrine. These two catecholamines are biosynthesized from tyrosine by a sequential process in which tyrosine is first hydroxylated to form dihydroxyphenylalanine, decarboxylated to dopamine, then β-hydroxylated to norepinephrine, and finally *N*-methylated to epinephrine.[1] Both chromaffin granules in the adrenal medulla and dense core vesicles in the sympathetic nerve endings contain large amount of epinephrine and norepinephrine. The secretion of these catecholamines is triggered by depolarization of postganglionic sympathetic neurons due to release of acetylcholine from preganglionic neurons. Epinephrine acts as both circulating hormone and neurotransmitter in the central nervous system, whereas norepinephrine mainly functions as a neurotransmitter rather than circulating hormone at normal physiological levels.[1,2]

1.2 The β_1- and β_2-ARs

The β-ARs were originally classified into β_1-AR and β_2-AR. During the past several decades, β_1-AR and β_2-AR have attracted much attention due to their clinical relevance in the therapy of several human diseases,[3] especially cardiovascular diseases. These two subtypes of β-ARs are encoded by two distinct intronless genes located on human chromosomes 10q24-26 and 5q31-32, respectively.[4–6] Extensive studies have clarified the physiological roles of these two receptors in mediating cardiovascular functions. In human heart, both β_1-AR and β_2-AR are involved in positive inotropic effects (heart contractility) and chronotropic effects (heart rate) despite the predominant expression of β_1-AR on the myocardium.[7,8] Both receptors couple to the stimulatory G-protein (Gs), which is responsible for adenylyl cyclase (AC) activation, subsequently inducing intracellular cAMP accumulation and protein kinase A (PKA) activation. The activation of Gs-cAMP-PKA pathway further leads to phosphorylation of L-type Ca^{2+} channels on the cell membrane, promoting Ca^{2+} influx and thus enhancing myocardial contraction. In addition, activation of Gs-cAMP-PKA pathway also induces phosphorylation of phospholamban, a 52-amino acid integral membrane protein responsible for Ca^{2+} uptake into the sarcoplasmic reticulum, and this process has been suggested to be involved in diastolic relaxation.[9] Moreover, β_2-AR is shown to also couple to the inhibitory G-protein (Gi), which counterbalances the stimulatory effects of Gs protein on cAMP production and activates the phosphatidyl inositol 3-kinase (PI3K)-Akt pathway. The switch of β_2-AR coupling from Gs to Gi in response to sustained stimulation has been known as the cause of several cardiovascular disorders, such as heart failure.[10] Apart from the cardiovascular effects, β_1-AR and β_2-AR are known to have effects on metabolism by accelerating substrate mobilization, such as lipolysis, glycogenolysis, and gluconeogenesis (reviewed in ref. 11).

2. The β_3-AR: Structure and characteristics

The genes encoding β_3-AR (*ADRB3*) have been identified and cloned from many species such as human, monkey, dog, bovine, mouse, rat, and guinea pig.[12] Distinct from the genes encoding β_1- and β_2-ARs, the genes encoding β_3-ARs contain introns, which raises the possibility that these genes may undergo alternative splicing, thus generating putative receptor isoforms with distinct pharmacological properties.[13] In human, the gene is located on chromosome 8p11-12.[14] There are two different structures of human *ADRB3* gene: the first structure with two exons and one intron[15–17] and the second structure with three exons and two introns.[18] The first structure, which is more widely accepted, encodes a protein of 408 amino acids (the first 402 amino acids from the first exon and the last 6 amino acids from the second exon) (Fig. 1). It has been suggested that two putative isoforms of human β_3-AR with different C-terminal tails (one with the first 402 amino acids and another with 6 additional amino acids on the C-terminal tail from the second exon) may exist after alternative splicing. In other species, such as dog and monkey, the structure of two exons and one intron has also been described.[19,20] In rats and mice, however, the *Adrb3* gene is known to contain three exons and two introns similar to the second structure in human *ADRB3*.[17,21]

β_3-AR differs from β_1- and β_2-ARs in pharmacological and regulatory properties. In pharmacological studies, some antagonists at β_1- and β_2-ARs, such as CGP12177A, bucindolol, and pindolol, are shown to be partial agonists for β_3-AR.[22–24] There are also several selective β_3-AR agonists including BRL37344, CL316243, and SR58611A, with high affinity and potency for β_3-AR but interact with β_1- and β_2-ARs only weakly.[25,26] The regulatory properties of β_3-AR are also distinct from β_1- and β_2-ARs. Both β_1- and β_2-ARs have phosphorylation sites (serine and threonine residues) in the C-terminal tails that act as substrates for G protein-coupled receptor kinase. However, the absence of phosphorylation sites on the C-terminus of β_3-AR suggests that this receptor may not undergo desensitization,[14] which is supported by the evidence from two studies that β_3-AR is resistant to short-term agonist-induced desensitization.[27,28] Moreover, β_3-AR is activated at higher concentrations of catecholamines

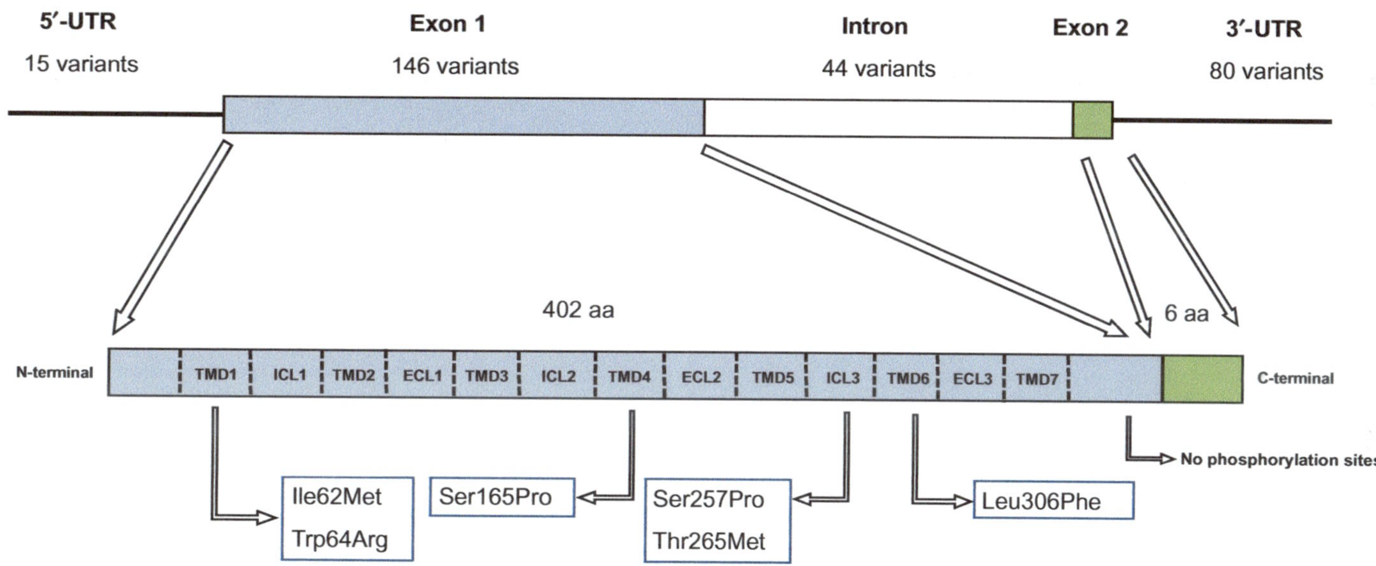

Fig. 1 The structure and polymorphic variants of human *ADRB3*. The human *ADRB3* encodes for a 408-amino acid protein with classical structure of GPCRs. The numbers of polymorphic variants with validation indicated in different regions were obtained from the Single Nucleotide Polymorphism dbSNP database. Nonsynonymous variants listed in coding region were investigated in clinical and/or functional studies.

than β_1- and β_2-ARs. Therefore, β_3-AR-mediated response can be preserved when the β_1- and β_2-AR-mediated responses are diminished.[29]

3. Physiological relevance of β_3-AR activation

Since its cloning in human (gene name *ADRB3*),[14] β_3-AR expression has been reported in several tissues such as white and brown adipose tissues, heart, blood vessels, gall bladder, gastrointestinal tract, prostate and urinary bladder detrusor, brain, and near-term myometrium (reviewed in ref. 30), suggesting that β_3-AR is an important regulator of various physiological functions.

3.1 The role of β_3-AR in mediating metabolic (Antiobesity and Antidiabetic) effects

The β_3-AR has attracted much attention on regulating metabolic processes since it was first discovered in rat adipose tissue. The physiological role of β_3-AR in adipocytes is well established in animal models. Activation of β_3-AR produces antiobesity effects as shown by several rodent studies. Yen et al. reported that selective β_3-AR ligand LY79771 stimulates metabolism and reduces body weight of obese mice, whereas it has no effect on body weight of normal mice.[31] Yoshida et al. investigated the effects of highly selective β_3-AR agonist CL316243 on obese mice and found that this compound can significantly increase resting metabolic rate and energy expenditure to generate negative balance and decrease body weight.[32]

Several studies have further clarified the mechanisms by which β_3-AR exerts its antiobesity effects in rodents.[33] The reduced body weight has been thought to be caused by increased energy expenditure mainly via thermogenesis in brown adipose tissue (BAT) with an increasing number of mitochondria and higher expression of uncoupling protein 1 (UCP-1).[33] UCP-1 is a mitochondrial inner membrane protein exclusive to BAT that functions as proton transporter, dissipating energy as heat and uncoupling oxidative phosphorylation without generation of ATP.[34] Moreover, fat oxidation in BAT and lipolysis in white adipose tissue (WAT) are also observed, suggesting that the antiobesity effects of β_3-AR activation are dependent on mobilization of fat from WAT depots to increase circulating free fatty acids (FFAs), which are in turn oxidized in BAT as the preferential source to support BAT thermogenesis.[33] Although some studies suggested that these antiobesity effects are due to increased energy expenditure without affecting food intake,[31,32] several other studies argued that antiobesity effects may be, at least in part, due to inhibition of food intake, with the evidence that acute treatment of rodents with β_3-AR selective agonists accounts for 40%–50% reduction in food intake and β_3-AR knockout mice are completely resistant to agonist stimulation on reducing food intake.[35–37]

The signaling pathways involved in these antiobesity effects are investigated by several studies in both cell systems and animal models, providing the molecular mechanism underlying these metabolic effects. The β_3-AR couples to both Gs and Gi proteins, leading to cAMP-PKA and c-Src-ERK1/2 pathways, respectively.[38,39] Cao et al. reported that Gs-cAMP-PKA pathway is responsible for thermogenesis response by sequentially activating p38 kinase and a subset of transcription factors for promoting UCP-1 protein expression in primary brown adipocytes.[40] On the other hand, Robidoux et al. demonstrated that Gs-cAMP-PKA pathway could contribute to lipolysis through phosphorylation of hormone-sensitive lipase. The Gi-c-Src-ERK1/2 pathway mediates lipolysis response only under the condition of maximal β_3-AR activation in response to agonist stimulation.[41]

In addition to antiobesity effects, β_3-AR activation is also shown to have antidiabetic effects in rodents. Acute treatment of mice with BRL26830A or CL316243 increases serum insulin levels by 10–100-fold.[36,42] The stimulatory effect of CL316243 on insulin secretion can be diminished when *Adrb3* is knocked out and completely rescued when it is transgenically reexpressed in white and/or brown adipocytes, further demonstrating that this response is mediated exclusively by β_3-AR.[35] In addition, Yoshida et al. showed that insulin sensitivity is improved after treatment of CL316243 on obese diabetic mice. These mice have increased expression of insulin receptors and decreased serum insulin and blood glucose levels during the glucose overloading test.[32] In rats, de Souza et al. and Liu et al. found that basal and insulin-stimulated glucose disposal are improved after β_3-AR stimulation by CL316243 or BRL37344.[43,44] The improved insulin sensitivity in rodents may account for these antidiabetic effects that are independent of the antiobesity effects and primary to β_3-AR activation.[45] It has been shown that WAT, BAT, and skeletal muscle are the tissues responsible for these β_3-AR-mediated antidiabetic effects[43,44] (reviewed in ref. 33).

Given the unequivocal metabolic effects mediated by β_3-AR activation in rodents, studies have been performed on human to investigate the therapeutic potential of β_3-AR agonists in treatment of obesity and diabetes. Although several clinical studies generated disappointing results that argue against the metabolic effects of β_3-AR activation in human since the receptor distribution pattern in human BAT and WAT is different from that in rodent BAT and WAT,[46,47] there are still several studies confirming that β_3-AR activation contributes to enhanced insulin action and fat oxidation in lean men and increased lipolysis and energy expenditure in overweight men.[48,49] It is also worth mentioning that some polymorphic variants in human *ADRB3* have been demonstrated to be associated with obesity and diabetes (discussed below), which highlights the important roles of β_3-AR in regulating such metabolic functions in human.

3.2 The role of β_3-AR in mediating cardiovascular effects

An increasing number of studies have also investigated the involvement of β_3-AR in physiological control of cardiovascular functions. In cardiovascular system, β_3-AR generally exerts negative inotropic effect in heart and vasorelaxation in blood vessels. The mechanisms of β_3-AR-mediated cardiovascular effects are potentially caused by alterations in excitation–contraction coupling and transmembrane ion channel activities.[50]

In different species, several studies have reported the negative inotropic effects induced by β_3-AR stimulation. For example, in beating guinea-pig heart and canine and rat cardiomyocytes, β_3-AR activation by BRL37344 produces the concentration-dependent negative inotropic effect associated with the inhibition of L-type Ca^{2+} channel and decrease of Ca^{2+} transient. The negative inotropic effect can be blocked by β_3-AR antagonists rather than β_1- and β_2-AR antagonists.[51–53] In rabbit ventricular cardiomyocytes, β_3-AR activation by several agonists produces a concentration-dependent negative inotropic effect associated with a reduction in Ca^{2+} transient and L-type Ca^{2+} current, and acceleration of action potential repolarization with an increase in delayed-rectifier K^+ current amplitude.[54]

In human, studies also confirmed the negative inotropic effect of β_3-AR stimulation. In contrast to the classical positive inotropic effect induced by isoproterenol alone, cotreatment of isoproterenol and nadolol (a β_1- and β_2-AR antagonist used to rule out the involvement of these two receptors) induces a negative inotropic effect in human endomyocardial biopsies.[55] The negative inotropic effect can also be observed in the presence of several β_3-AR agonists including BRL37344, SR58611A, and CL316243,[55] as well as the naturally occurring catecholamine, norepinephrine, at concentrations close to pathophysiologically relevant levels.[56] However, the regulation of contractility by β_3-AR stimulation in human atria is different from that in human ventricles since functional studies using β_3-AR agonists have generated conflicting results in atria tissue.[57,58]

In blood vessels, the involvement of β_3-AR in vasorelaxation has been suggested by several studies despite that this effect is mediated essentially through β_2-AR activation and slightly through β_1-AR activation. In rat aorta, the main β-AR subtype is demonstrated to be β_3-AR, followed by β_1-AR with less expression of β_2-AR.[59] In rat isolated common carotid arteries and thoracic aortas, β_3-AR agonists produce significant relaxation independent of β_1- and β_2-AR activation.[60–62] The vasorelaxation induced by β_3-AR stimulation was also confirmed in vivo. In conscious dogs, treatment of β_3-AR agonists leads to peripheral vasorelaxation primarily in skin and adipose tissue.[63,64] Similarly, in unanesthetized rats, β_3-AR agonists also produce vasorelaxation with lesser extent than in dogs, which mainly occurs in BAT.[65,66]

The signaling pathways trigged by β_3-AR in cardiovascular tissues have been elucidated. In cardiomyocytes, β_3-AR couples to Gi protein, leading to activation of endothelial nitric oxide synthase (eNOS). Nitric oxide (NO) is then produced to activate the soluble isoform of guanylyl cyclase that increases the concentration of cGMP, further activating protein kinase G and L-type Ca^{2+} channels, to exert negative inotropic effect. In blood vessels, β_3-AR-mediated vasorelaxation is dependent on both Gi-eNOS-cGMP and Gs-cAMP-PKA pathways, which activates K^+ channels and L-type Ca^{2+} channels, respectively (reviewed in refs. 29,30).

The β_3-AR is also widely distributed in nonvascular smooth muscle in different tissues, such as gastrointestinal tract, urinary bladder detrusor, and myometrium. The general effect of β_3-AR activation in these tissues is suggested to be relaxation of smooth muscle, leading to increased compliance of gut and gastric volume, attenuated bladder contraction during the storage phase of the micturition cycle, and attenuated myometrium contraction especially at the end of the pregnancy (reviewed in ref. 30).

4. Human *ADRB3* polymorphic variants: Potential pathophysiological roles

The Single Nucleotide Polymorphism database (dbSNP) (https://www.ncbi.nlm.nih.gov/snp) provides more than 1000 polymorphisms of human *ADRB3*. We excluded the variants without validation, and obtained 285 variants, with 15 in 5′-untranslated region (UTR), 80 in 3′-UTR, 44 in intron, and 146 in exon (Fig. 1). SNPs in the 5′- and 3′-UTR might change gene expression level. Missense variants in the exons might cause a loss of function in the receptor by different mechanisms such as impaired protein production, plasma membrane expression, ligand binding, or signaling.[67,68] Currently, there are 13 SNPs (validated by 1000 Genomes, cluster, and frequency) in coding sequence from the dbSNP database (Table 1), including missense (rs201940081, rs200142610, rs4994, rs184077488, rs150886462, rs199753604, rs142082800, rs200163984, rs148637032, rs4995, rs374271379, and rs36031925) and frame-shift (rs569810926) mutations.

4.1 Prevalence of *ADRB3* polymorphic variants and clinical features of carriers

Of these variants, the most common *ADRB3* variant is SNP rs4994 (c.190T > C or p.Trp64Arg), resulting in a tryptophan to arginine exchange at position 64 in the intracellular end of transmembrane domain 1 (TMD1). Trp64Arg variant has attracted significant scientific interest with regard to its potential pathogenetic role in metabolic disorders, such as obesity, type 2 diabetes mellitus (T2DM), and related disorders. Several independent groups simultaneously reported that homozygous Arg substitution at Trp64 affects the metabolic functions by lowering the metabolic rate, increasing the risk for obesity and insulin resistance, and lowering age of onset of T2DM.[71–74] The Trp64Arg polymorphism is present in nearly all populations of the world with an overall allelic frequency of 10.57% (Table 1).

A substantial number of studies confirm the association of Trp64Arg polymorphism with disorders in different populations. Walston et al. detected this SNP with allelic frequency of 31% in Pima Indians.[73] The association study was performed in 642 Pima subjects (390 with T2DM and 252 without T2DM), showing that the mean age at the onset of T2DM of subjects homozygous for Arg64 (36 ± 10 years) is significantly lower than those who are Trp64Arg heterozygotes (40 ± 10 years) or normal Trp64 homozygotes (41 ± 11 years). Furthermore, subjects with this SNP are susceptible to have a lower adjusted resting metabolic rate.[73] Kurabayashi et al. investigated the relationship of Trp64Arg with obesity-related traits in 686 elderly subjects (429 women and 257 men) from a cross section of a normal population in Australia.[75] The allelic frequency is 7.5%, and female heterozygotes exhibit significantly increased total body weight (67.5 ± 12.9 vs 64.1 ± 12.2 kg) and body mass index (BMI) (26.3 ± 4.7 vs 25.1 ± 4.5 kg/m^2). Meanwhile, they have a longer reproductive life, with an earlier menarche, a higher gravidity, and higher parity.[75]

In Japanese, this SNP is detected with allelic frequency of 19.0% and many subjects with Trp64Arg variant exhibit abnormal obesity and/or

Table 1 Nonsynonymous variants in the coding region of human *ADRB3* (Validated by 1000 genomes, cluster, and frequency).

dbSNP	Allele/Residue Change		Allele Count	Allele Number	Number of Homozygotes	Allelic Frequency
	Allele	Residue				
rs201940081	c.12G>C	p.Trp4Cys	4	61670	0	0.00006486
rs200142610	c.49C>A	p.Leu17Ile	Not available			
rs4994	c.190T>C	p.Trp64Arg	12257	115984	797	0.1057
rs184077488	c.461C>A	p.Ala154Asp	12	27166	0	0.0004417
rs150886462	c.512C>G	p.Ala171Gly	14	20618	0	0.0006790
rs199753604	c.557C>A	p.Ala186Glu	5	19690	0	0.0002539
rs142082800	c.570C>G	p.His190Gln	Not available			
rs200163984	c.193C>T	p.Pro193Leu	3	20588	0	0.0001457
rs148637032	c.673G>A	p.Ala225Thr	8	19912	0	0.0004018
rs4995	c.794C>T	p.Thr265Met	13	10470	0	0.001242
rs374271379	c.1033A>G	p.Ile345Val	3	16148	0	0.0001858
rs36031925	c.1057C>T	p.Arg353Cys	69	15204	1	0.004538
rs569810926	c.731delGC	p.Arg244fs	Not available			

Data are derived from the Single Nucleotide Polymorphism dbSNP database (https://www.ncbi.nlm.nih.gov/snp)[69] and the Exome Aggregation Consortium (ExAC) (http://exac.broadinstitute.org/).[70]

diabetes-related phenotypes, such as higher BMI, hyperinsulinemia, and/or earlier onset of T2DM.[72,76–78] In Chinese, this SNP has been identified with an allelic frequency of 12.0%.[79] Thomas et al. investigated 802 Southern Chinese subjects to support a direct involvement of Trp64Arg in development of obesity but not diabetes and dyslipidemia.[80] Mo et al. also reported that Trp64Arg variant is associated with obesity in Chinese male hypertensive patients with a higher frequency in obese hypertensive patients than in the nonobese hypertensive population and healthy controls.[81] In Europeans, Trp64Arg SNP is observed in French, Danes, Germans, British, Swedes, and Finns, with the allelic frequencies ranging from 4.7% in French to 10.5% in Finns (reviewed in ref. 82). Several studies also confirmed the pathophysiological role of Trp64Arg variant in development of obesity and/or diabetes in these populations.

In addition to the association with obesity and diabetes, the Trp64Arg variant is also shown to be associated with development of other disorders, such as cardiovascular disease, gallstone formation, hyperuricemia, and overactive bladder syndrome. Ringel et al. reported that the variant contributes to hypertension or vascular complications in patients with T2DM.[83] Xifra et al. investigated 106 subjects from Northern Spain to propose that the presence of Trp64Arg variant is associated not only with endothelial dysfunction but also with significant paradoxical vasoconstriction.[84] One recent study based on meta-analysis also reveals the role of Trp64Arg variant in development of essential hypertension, especially in Chinese and Caucasian populations.[85] Klass et al. identified Trp64Arg variant as a genetic marker for gallstone formation in Germans.[86] They found a higher frequency of the Arg64 allele in gallstone carriers compared to age-, gender-, and BMI-matched controls (5.9% vs 0.7%). Individuals carrying this variant have more frequently first-degree relatives with gallstone disease, further suggesting the genetic importance of Trp64Arg.[86] Moreover, studies from three different populations, Korean, Chinese, and Spanish, reported the correlation of Trp64Arg with risk of hyperuricemia.[87–89] Two studies showed that this variant is significantly associated with overactive bladder syndrome in Japanese and Chinese populations.[89,90]

Although numerous studies have been performed on the correlation between the Trp64Arg polymorphism of *ADRB3* and various disorders, there are also many studies refuting this potential relationship due to no significant differences observed between subjects with Trp64Arg variant and their controls (reviewed in ref. 91). The existence of divergent results may be attributable to the number of subjects and ethnic origin of the population studied, as well as interactions of Trp64Arg variant with environment or variants in other genes. Kurokawa et al. performed a meta-analysis of 97 studies involving 44,833 individuals to assess the association between Trp64Arg and weight gain. In their study, Trp64Arg variant is shown to be associated with weight gain in East Asians rather than Europeans, suggesting an ethnicity-dependent susceptibility.[92]

It also seems that there is a strong linkage between the presence of Trp64Arg variant and other polymorphic variants of the same gene or other genes. For example, Hoffstedt et al. and Teitsma et al. reported that Trp64Arg is in linkage to several noncoding variants in the *ADRB3* intronic region or 3′-UTR and 3′-downstream regions, such as 1219A > T, 2135A > G, 2502G > C, and 3558G > T, suggesting the existence of the haplotype blocks that possibly affect transcription of the gene and/or the stability of the mRNA transcript, thereby influencing receptor expression and contributing to the phenotypes.[93,94] Carlsson et al. genotyped 1054 nondiabetic and diabetic Caucasians, revealing the linkage between *ADRB2* (encoding β_2-AR) Gln27Glu variant and *ADRB3* Trp64Arg variant.[95] Furthermore, several studies reported the interactions of *ADRB3* Trp64Arg variant with SNPs in other genes, such as *type 2 deiodinase, lipoprotein lipase*, and *NADH dehydrogenase subunit-2 genes*,[96–98] which are crucial in lipid and energy metabolism. Therefore, it is possible that Trp64Arg does not impair receptor function by itself to an extent that is detrimental to the physiological functions and phenotypes.

In addition to Trp64Arg, there are also several other SNPs and mutations identified in the coding or regulatory regions of *ADRB3*, which are thought to be associated with human diseases as supported by clinical studies. Halushka et al. reported a very rare SNP rs4995 (c.794C > T or p.Thr265Met) in the intracellular loop 3 (ICL3) of the receptor, which may be in linkage disequilibrium with Trp64Arg.[99] Vrydag et al. and Teitsma et al. detected a SNP, c.918G > C or p.Leu306Phe, in the TMD6, residing directly next to an amino acid which is conserved in all nine ARs and involved in catecholamine binding.[94,100] In the study based on a collection of large extended pedigrees of Mexican American descent, Lehman et al. identified a novel mutation, Ile62Met (c.186C > G), in the carboxy end of TMD1.[101] This mutation was shown to increase the risk for T2DM and reduce the age-of-diabetes onset, although it does not affect obesity risk. The amino acid Ile62 is evolutionarily conserved in vertebrates, suggesting that substitution on this residue may affect receptor function.[101]

Recently, Huang et al. genotyped a total of 1337 healthy control subjects and unrelated patients with T2DM in Chinese population, and identified two other SNPs in the coding region including rs72655364 (c.493T>C or p.Ser165Pro) and rs72655365 (c.769T>C or p.Ser257Pro).[88] These two SNPs are present in diabetic patients with significantly higher frequencies compared to healthy controls. Patients with Ser165Pro variant exhibit higher hemoglobin A1c, fasting plasma glucose, and postprandial plasma glucose levels than Ser165 genotype subjects. Also, lower fasting and postprandial serum insulin levels and insulin resistance are observed in patients with Ser257Pro variant.[88] Furthermore, in vitro functional study demonstrated that these two mutations are detrimental to receptor function, which may account for their pathophysiological roles in susceptibility and development of T2DM.[88] For SNPs in noncoding regions, Voruganti et al. reported that SNP rs35361594 in the 5′-UTR of *ADRB3* may play a pathophysiological role in the development of obesity in Alaskan Eskimos, since this SNP shows an association with BMI, percentage body fat, and waist circumference.[102] Hamilton et al. studied SNP rs4998 in the 3′-UTR of *ADRB3* and found it to be associated with Alzheimer's disease in ε4 allele of the Apolipoprotein E gene (ApoE4)-negative samples (ApoE4 is a polymorphism strongly associated with Alzheimer's disease).[103]

4.2 Functional characterizations of β_3-AR variants

The impact of Trp64Arg variant on metabolic functions of β_3-AR has been assessed by several in vitro and ex vivo functional studies which involve investigations on receptor–ligand binding properties, signaling properties (Gs-cAMP signaling pathway), receptor desensitization, and lipolytic activity of transfected cells or isolated adipocytes.

In vitro, three studies clarified that Trp64Arg variant affects the receptor function, resulting in impaired ligand-induced cAMP accumulation or ligand-induced insulin secretion. Piétri-Rouxel et al. found that in CHO-K1 and HEK293 cells, mutant receptor results in significantly reduced maximal cAMP accumulation independent of Gi coupling under stimulation of various β_3-AR agonists, including endogenous catecholamines (epinephrine and norepinephrine), nonselective agonist isoproterenol, and selective agonist CGP12177A, although the binding properties of mutant receptor are not affected.[104] Reduced basal cAMP accumulation (constitutive activity) is also observed in CHO-K1 cells transfected with mutant receptor.[104] Similar results were also obtained by Kimura et al. through investigation of the signaling properties of mutant receptor in mouse 3T3-L1 cells. Compared with wild-type receptor, the mutant receptor is less responsive to endogenous catecholamines, isoproterenol, and another selective agonist L-755507, with lower maximal cAMP accumulation, although EC_{50} value for cAMP accumulation is decreased.[105] Perfetti et al. observed abnormal insulin secretion in rat insulinoma (RIN) cells expressing mutant receptor in response to selective agonists CL3162435 and BRL37344.[106]

In contrast to the results demonstrating that Trp64Arg variant impairs receptor function, two groups, Candelore et al. and Vrydag et al., respectively, reported that there are no differences in ligand binding properties, cAMP accumulation, and receptor desensitization between Trp64Arg mutant and wild-type receptor upon stimulation with various agonists in CHO and HEK293 cells.[100,107] The Trp64Arg variant in combination with two other variants, Thr265Met or Leu306Phe, previously shown to be in linkage with Trp64Arg, also has little effect on ligand binding, function, or regulation of receptor responsiveness.[100] Moreover, Isogaya et al. observed enhanced cAMP response when mutant receptor is coexpressed with the AC type III isoform and stimulated by CGP12177A and isoproterenol in COS-6 cells. The binding and signaling properties of wild-type and mutant receptors are indistinguishable in COS-6 cells lacking AC type III.[108]

Ex vivo, Li et al. found no distinguishable lipolytic activities in response to norepinephrine or CGP12177 between isolated visceral adipocytes obtained from heterozygous Trp64Arg and homozygous Trp64 subjects.[109] Although Hoffstedt et al. also observed no difference between visceral adipocytes obtained from the homozygous Arg64 and homozygous Trp64 subjects regarding either basal or ligand-induced maximal lipolysis, the sensitivity of the lipolytic response for CGP12177 is 10-fold lower in subjects carrying Trp64Arg variant.[93] In addition, Umekawa et al. found that the β_3-AR selective agonist L-755507 rather than isoproterenol or CGP12177 induces significantly lowered lipolytic activity, including lowered potency and efficacy, in omental adipocytes from homozygous Arg64 subjects.[110]

Additionally, Huang et al. performed functional studies of two other variants Ser165Pro and Ser257Pro associated with T2DM in Chinese population.[88] In both HEK293 and CHO-K1 cells, these two mutations do not affect receptor expression and trafficking but impair cAMP accumulation in response to different concentrations of CL316243. Proline has been known as a key residue in protein folding due to its unique structure,[111] and serine at position 257 in ICL3 is considered as a putative phosphorylation site. It is reasonable to speculate that proline substitution on position 165 or 257 might impair receptor structure and function, leading to impaired cAMP signaling. In this study, ERK1/2 phosphorylation induced by CL316243 is not affected by Ser165Pro and Ser257Pro in HEK293 cells. The β_3-AR is known to couple to Gi protein to induce ERK1/2 phosphorylation.[38,39] It seems that Ser165Pro and Ser257Pro preferentially affect interaction of receptor with Gs protein rather than Gi protein.[88]

It should be noted that data from the current functional studies are still limited for assessment of pathophysiological roles of these variants with regard to other physiological functions, such as cardiovascular effects, because there are currently no functional studies available for investigation of other distinct signaling pathways such as activation of different types of Ca^{2+} and K^+ channels or NO synthase, apart from cAMP accumulation or ERK1/2 activation.

5. Summary

During the past three decades, extensive genetic and pharmacological studies have elucidated the important roles of the β_3-AR in mediating distinct physiological functions in various types of cells and tissues, such as adipose tissue and cardiovascular system, as well as gastrointestinal tract, urinary bladder, etc. In addition, polymorphic variants of *ADRB3* are demonstrated to be associated with obesity and diabetes, as well as other disorders. However, there are also many studies arguing against this potential association due to conflicting results obtained from the same or different populations. It seems that these *ADRB3* polymorphisms are disease-risk factors rather than disease-causing factors, which might influence the progression of such diseases.[112] In order to further explore the relationship between these genetic variants and diseases, more comprehensive functional studies on additional signaling pathways triggered by the receptor are needed.

Acknowledgements

L-KY was supported by stipend from China Scholarship Council, People's Republic of China.

References

1. Landsberg L. *Catecholamines. Pheochromocytomas, Paragangliomas and Disorders of the Sympathoadrenal System*. Springer; 2018:1–14.
2. Young JB, Landsberg L. Chapter 13. Physiology of the sympathoadrenal system. In: *WilliamsTextbook of Endocrinology*. 9th ed. Philadelphia: Saunders; 1998.
3. Stiles GL, Caron MG, Lefkowitz RJ. β-Adrenergic receptors: biochemical mechanisms of physiological regulation. *Physiol Rev*. 1984;64:661–743.
4. Dixon RAF, Kobilka BK, Strader DJ, et al. Cloning of the gene and cDNA for mammalian β-adrenergic receptor and homology with rhodopsin. *Nature*. 1986;321:75–79.

5. Kobilka BK, Dixon RA, Frielle T, et al. cDNA for the human β_2-adrenergic receptor: a protein with multiple membrane-spanning domains and encoded by a gene whose chromosomal location is shared with that of the receptor for platelet-derived growth factor. *Proc Natl Acad Sci USA*. 1987;84:46–50.
6. Yang-Feng TL, Xue FY, Zhong WW, et al. Chromosomal organization of adrenergic receptor genes. *Proc Natl Acad Sci USA*. 1990;87:1516–1520.
7. McDevitt DG. In vivo studies on the function of cardiac β-adrenoceptors in man. *Eur Heart J*. 1989;10:22–28.
8. Santos IN, Spadari-Bratfisch RC. Stress and cardiac beta adrenoceptors. *Stress*. 2006;9:69–84.
9. Brodde OE, Michel MC. Adrenergic and muscarinic receptors in the human heart. *Pharmacol Rev*. 1999;51:651–690.
10. Spadari RC, Cavadas C, de Carvalho AETS, Ortolani D, de Moura AL, Vassalo PF. Role of beta-adrenergic receptors and sirtuin signaling in the heart during aging, heart failure, and adaptation to stress. *Cell Mol Neurobiol*. 2018;38:109–120.
11. Coppack SW, Jensen MD, Miles JM. In vivo regulation of lipolysis in humans. *J Lipid Res*. 1994;35:177–193.
12. Strosberg AD. Structure and function of the β_3-adrenergic receptor. *Annu Rev Pharmacol Toxicol*. 1997;37:421–450.
13. Rozec B, Gauthier C. β_3-Adrenoceptors in the cardiovascular system: putative roles in human pathologies. *Pharmacol Ther*. 2006;111:652–673.
14. Emorine LJ, Marullo S, Briend-Sutren MM, et al. Molecular characterization of the human beta 3-adrenergic receptor. *Science*. 1989;245:1118–1121.
15. Granneman JG, Lahners KN, Rao DD. Rodent and human β_3-adrenergic receptor genes contain an intron within the protein-coding block. *Mol Pharmacol*. 1992;42: 964–970.
16. Granneman JG, Lahners KN, Chaudhry A. Characterization of the human β_3-adrenergic receptor gene. *Mol Pharmacol*. 1993;44:264–270.
17. Van Spronsen A, Nahmias C, Krief S, Briend-Sutren MM, Strosberg AD, Emorine LJ. The promoter and intron/exon structure of the human and mouse β3-adrenergic-receptor genes. *Eur J Biochem*. 1993;213:1117–1124.
18. Levasseur S, Pigeon C, Reyl-Desmars F, Caput D, Lewin MJ. Stimulation of adenylate cyclase by the isoforms of human and rat beta-3 adrenergic receptor expressed in the CHO cells. *Gastroenterol Clin Biol*. 1995;19:668–672.
19. Walston J, Lowe A, Silver K, et al. The β_3-adrenergic receptor in the obesity and diabetes prone rhesus monkey is very similar to human and contains arginine at codon 64. *Gene*. 1997;188:207–213.
20. Lenzen G, Pietri-Rouxel F, Drumare MF, et al. Genomic cloning and species-specific properties of the recombinant canine β_3-adrenoceptor. *Eur J Pharmacol*. 1998;363: 217–227.
21. Evans BA, Papaioannou M, Hamilton S, Summers RJ. Alternative splicing generates two isoforms of the β_3-adrenoceptor which are differentially expressed in mouse tissues. *Br J Pharmacol*. 1999;127:1525–1531.
22. Liggett SB. Functional properties of the rat and human β_3-adrenergic receptors: differential agonist activation of recombinant receptors in Chinese hamster ovary cells. *Mol Pharmacol*. 1992;42:634–637.
23. Blin N, Camoin L, Maigret B, Strosberg AD. Structural and conformational features determining selective signal transduction in the β_3-adrenergic receptor. *Mol Pharmacol*. 1993;44:1094–1104.
24. Kaumann AJ, Molenaar P. Differences between the third cardiac β-adrenoceptor and the colonic β3-adrenoceptor in the rat. *Br J Pharmacol*. 1996;118:2085–2098.

25. Emorine L, Blin N, Strosberg AD. The human β3-adrenoceptor: the search for a physiological function. *Trends Pharmacol Sci.* 1994;15:3–7.
26. Strosberg AD, Pietri-Rouxel F. Function and regulation of the β3-adrenoceptor. *Trends Pharmacol Sci.* 1996;17:373–381.
27. Liggett SB, Freedman NJ, Schwinn DA, Lefkowitz RJ. Structural basis for receptor subtype-specific regulation revealed by a chimeric β3/β2-adrenergic receptor. *Proc Natl Acad Sci USA.* 1993;90:3665–3669.
28. Nantel F, Bonin H, Emorine LJ, et al. The human β3-adrenergic receptor is resistant to short term agonist-promoted desensitization. *Mol Pharmacol.* 1993;43:548–555.
29. Gauthier C, Langin D, Balligand JL. β3-Adrenoceptors in the cardiovascular system. *Trends Pharmacol Sci.* 2000;21:426–431.
30. Ursino MG, Vasina V, Raschi E, Crema F, De Ponti F. The β3-adrenoceptor as a therapeutic target: current perspectives. *Pharmacol Res.* 2009;59:221–234.
31. Yen TT. The antiobesity and metabolic activities of LY79771 in obese and normal mice. *Int J Obes.* 1984;8:69–78.
32. Yoshida T, Sakane N, Wakabayashi Y, Umekawa T, Kondo M. Anti-obesity effect of CL 316,243, a highly specific β3-adrenoceptor agonist, in mice with monosodium-L-glutamate-induced obesity. *Eur J Endocrinol.* 1994;131:97–102.
33. Weyer C, Gautier JF, Danforth Jr E. Development of beta 3-adrenoceptor agonists for the treatment of obesity and diabetes—an update. *Diabetes Metab.* 1999;25:11–21.
34. Palou A, Picó C, Bonet ML, Oliver P. The uncoupling protein, thermogenin. *Int J Biochem Cell Biol.* 1998;30:7–11.
35. Grujic D, Susulic VS, Harper ME, et al. β3-Adrenergic receptors on white and brown adipocytes mediate β3-selective agonist-induced effects on energy expenditure, insulin secretion, and food intake. A study using transgenic and gene knockout mice. *J Biol Chem.* 1997;272:17686–17693.
36. Yoshida T. The antidiabetic β3-adrenoceptor agonist BRL 26830A works by release of endogenous insulin. *Am J Clin Nutr.* 1992;55:237S–241S.
37. Tsujii S, Bray GA. Food intake of lean and obese Zucker rats following ventricular infusions of adrenergic agonists. *Brain Res.* 1992;587:226–232.
38. Cao W, Luttrell LM, Medvedev AV, et al. Direct binding of activated c-Src to the β3-adrenergic receptor is required for MAP kinase activation. *J Biol Chem.* 2000;275:38131–38134.
39. Soeder KJ, Snedden SK, Cao W, et al. The β3-adrenergic receptor activates mitogen-activated protein kinase in adipocytes through a Gi-dependent mechanism. *J Biol Chem.* 1999;274:12017–12022.
40. Cao W, Medvedev AV, Daniel KW, Collins S. β-Adrenergic activation of p38 MAP kinase in adipocytes: cAMP induction of the uncoupling protein 1 (UCP1) gene requires p38 MAP kinase. *J Biol Chem.* 2001;276:27077–27082.
41. Robidoux J, Kumar N, Daniel KW, et al. Maximal β3-adrenergic regulation of lipolysis involves Src and epidermal growth factor receptor-dependent ERK1/2 activation. *J Biol Chem.* 2006;281:37794–37802.
42. Susulic VS, Frederich RC, Lawitts J, et al. Targeted disruption of the β3-adrenergic receptor gene. *J Biol Chem.* 1995;270:29483–29492.
43. de Souza CJ, Hirshman MF, Horton ES. CL-316,243, a β3-specific adrenoceptor agonist, enhances insulin-stimulated glucose disposal in nonobese rats. *Diabetes.* 1997;46:1257–1263.
44. Liu YL, Cawthorne MA, Stock MJ. Biphasic effects of the β-adrenoceptor agonist, BRL 37344, on glucose utilization in rat isolated skeletal muscle. *Br J Pharmacol.* 1996;117:1355–1361.
45. Arch JR, Wilson S. Prospects for beta 3-adrenoceptor agonists in the treatment of obesity and diabetes. *Int J Obes Relat Metab Disord.* 1996;20:191–199.

46. Barbe P, Millet L, Galitzky J, Lafontan M, Berlan M. In situ assessment of the role of the β1, β2-and β3-adrenoceptors in the control of lipolysis and nutritive blood flow in human subcutaneous adipose tissue. *Br J Pharmacol.* 1996;117:907–913.
47. Deng C, Paoloni-Giacobino A, Kuehne F, et al. Respective degree of expression of β1, β2-and β3-adrenoceptors in human brown and white adipose tissues. *Br J Pharmacol.* 1996;118:929–934.
48. Weyer C, Tataranni PA, Snitker S, Danforth E, Ravussin E. Increase in insulin action and fat oxidation after treatment with CL 316,243, a highly selective beta3-adrenoceptor agonist in humans. *Diabetes.* 1998;47:1555–1561.
49. van Baak MA, Hul GB, Toubro S, et al. Acute effect of L-796568, a novel β3-adrenergic receptor agonist, on energy expenditure in obese men. *Clin Pharmacol Ther.* 2002;71:272–279.
50. Gauthier C, Rozec B, Manoury B, Balligand JL. Beta-3 adrenoceptors as new therapeutic targets for cardiovascular pathologies. *Curr Heart Fail Rep.* 2011;8:184–192.
51. Kitamura T, Onishi K, Dohi K, Okinaka T, Isaka N, Nakano T. The negative inotropic effect of β3-adrenoceptor stimulation in the beating guinea pig heart. *J Cardiovasc Pharmacol.* 2000;35:786–790.
52. Sane DC. Upregulation of functional β3-adrenergic receptor in the failing canine myocardium. *Circ Res.* 2001;89:599–606.
53. Zhang ZS, Cheng HJ, Onishi K, Ohte N, Wannenburg T, Cheng CP. Enhanced inhibition of L-type Ca^{2+} current by β3-adrenergic stimulation in failing rat heart. *J Pharmacol Exp Ther.* 2005;315:1203–1211.
54. Audigane L, Kerfant BG, El Harchi A, et al. Rabbit, a relevant model for the study of cardiac β3-adrenoceptors. *Exp Physiol.* 2009;94:400–411.
55. Gauthier C, Tavernier G, Charpentier F, Langin D, Le Marec H. Functional β_3-adrenoceptor in the human heart. *J Clin Invest.* 1996;98:556–562.
56. Gauthier C, Leblais V, Kobzik L, et al. The negative inotropic effect of β_3-adrenoceptor stimulation is mediated by activation of a nitric oxide synthase pathway in human ventricle. *J Clin Invest.* 1998;102:1377–1384.
57. Kaumann AJ, Molenaar P. Modulation of human cardiac function through 4 β-adrenoceptor populations. *Naunyn Schmiedebergs Arch Pharmacol.* 1997;355:667–681.
58. Skeberdis VA, Gendvilienė V, Zablockaitė D, et al. β3-Adrenergic receptor activation increases human atrial tissue contractility and stimulates the L-type Ca^{2+} current. *J Clin Invest.* 2008;118:3219–3227.
59. Oliver E, Martí D, Montó F, et al. The impact of α1-adrenoceptors up-regulation accompanied by the impairment of β-adrenergic vasodilatation in hypertension. *J Pharmacol Exp Ther.* 2009;328:982–990.
60. Oriowo MA. Atypical β-adrenoceptors in the rat isolated common carotid artery. *Br J Pharmacol.* 1994;113:699–702.
61. Trochu JN, Leblais V, Rautureau Y, et al. Beta 3-adrenoceptor stimulation induces vasorelaxation mediated essentially by endothelium-derived nitric oxide in rat thoracic aorta. *Br J Pharmacol.* 1999;128:69–76.
62. Oriowo MA. Different atypical β-adrenoceptors mediate isoprenalineinduced relaxation in vascular and non-vascular smooth muscles. *Life Sci.* 1995;56:PL269–PL275.
63. Shen YT, Zhang H, Vatner SF. Peripheral vascular effects of beta-3 adrenergic receptor stimulation in conscious dogs. *J Pharmacol Exp Ther.* 1994;268:466–473.
64. Berlan M, Galitzky J, Bousquet-Melou A, Lafontan M, Montastruc JL. Beta-3 adrenoceptor-mediated increase in cutaneous blood flow in the dog. *J Pharmacol Exp Ther.* 1994;268:1444–1451.
65. Shen YT, Cervoni P, Claus T, Vatner SF. Differences in β_3-adrenergic receptor cardiovascular regulation in conscious primates, rats and dogs. *J Pharmacol Exp Ther.* 1996;278:1435–1443.

66. Takahashi H, Yoshida T, Nishimura M, Nakanishi T, Kondo M, Yoshimura M. Beta-3 adrenergic agonist, BRL-26830A, and alpha/beta blocker, arotinolol, markedly increase regional blood flow in the brown adipose tissue in anesthetized rats. *Jpn Circ J.* 1992;56:936–942.
67. Tao YX. Inactivating mutations of G protein-coupled receptors and diseases: structure-function insights and therapeutic implications. *Pharmacol Ther.* 2006;111:949–973.
68. Tao YX, Conn PM. Chaperoning G protein-coupled receptors: from cell biology to therapeutics. *Endocr Rev.* 2014;35:602–647.
69. Sherry ST, Ward MH, Kholodov M, et al. dbSNP: the NCBI database of genetic variation. *Nucleic Acids Res.* 2001;29:308–311.
70. Karczewski KJ, Weisburd B, Thomas B, et al. The ExAC browser: displaying reference data information from over 60 000 exomes. *Nucleic Acids Res.* 2016;45:D840–D845.
71. Clément K, Vaisse C, Manning BSJ, et al. Genetic variation in the β3-adrenergic receptor and an increased capacity to gain weight in patients with morbid obesity. *N Engl J Med.* 1995;333:352–354.
72. Kadowaki H, Yasuda K, Iwamoto K, et al. A mutation in the β3-adrenergic receptor gene is associated with obesity and hyperinsulinemia in Japanese subjects. *Biochem Biophys Res Commun.* 1995;215:555–560.
73. Walston J, Silver K, Bogardus C, et al. Time of onset of non-insulin-dependent diabetes mellitus and genetic variation in the β3-adrenergic–receptor gene. *N Engl J Med.* 1995;333:343–347.
74. Widen E, Lehto M, Kanninen T, Walston J, Shuldiner AR, Groop LC. Association of a polymorphism in the β3-adrenergic–receptor gene with features of the insulin resistance syndrome in Finns. *N Engl J Med.* 1995;333:348–352.
75. Kurabayashi T, Carey DGP, Morrison NA. The β3-adrenergic receptor gene Trp64Arg mutation is overrepresented in obese women: effects on weight, BMI, abdominal fat, blood pressure, and reproductive history in an elderly Australian population. *Diabetes.* 1996;45:1358–1363.
76. Kotani K, Sakane N, Kurozawa Y, et al. Polymorphism of Trp64Arg in β3-adrenergic receptor gene and serum LDL-cholesterol concentrations in healthy Japanese. *Ann Clin Biochem.* 2008;45:313–315.
77. Takeuchi S, Katoh T, Yamauchi T, Kuroda Y. ADRB3 polymorphism associated with BMI gain in Japanese men. *Exp Diabetes Res.* 2012;2012:973561.
78. Yamakita M, Ando D, Tang S, Yamagata Z. The Trp64Arg polymorphism of the β3-adrenergic receptor gene is associated with weight changes in obese Japanese men: a 4-year follow-up study. *J Physiol Anthropol.* 2010;29:133–139.
79. Silver K, Walston J, Wang Y, Dowse G, Zimmet P, Shuldiner AR. Molecular scanning for mutations in the β_3-adrenergic receptor gene in Nauruans with obesity and noninsulin-dependent diabetes mellitus. *J Clin Endocrinol Metab.* 1996;81:4155–4158.
80. Thomas GN, Tomlinson B, Chan JCN, Young RP, Critchley JA. The Trp64Arg polymorphism of the β3-adrenergic receptor gene and obesity in Chinese subjects with components of the metabolic syndrome. *Int J Obes Relat Metab Disord.* 2000;24: 545–551.
81. Mo W, Zhang GG, Yang TL, et al. The genetic polymorphisms of β3-adrenergic receptor (AR) Trp64Arg and β2-AR Gln27Glu are associated with obesity in Chinese male hypertensive patients. *Clin Chem Lab Med.* 2007;45:493–498.
82. Strosberg AD. Association of β_3-adrenoceptor polymorphism with obesity and diabetes: current status. *Trends Pharmacol Sci.* 1997;18:449–454.
83. Ringel J, Kreutz R, Distler A, Sharma AM. The Trp64Arg polymorphism of the β3-adrenergic receptor gene is associated with hypertension in men with type 2 diabetes mellitus. *Am J Hypertens.* 2000;13:1027–1031.

84. Xifra G, Castro A, Ortega FJ, Ricart W, Fernández-Real JM. The Trp64Arg β3-adrenergic receptor gene polymorphism is associated with endothelium-dependent vasodilatation. *J Hum Hypertens*. 2015;29:134.
85. Li YY, Lu XZ, Wang H, et al. ADRB3 gene Trp64Arg polymorphism and essential hypertension: a meta-analysis including 9,555 subjects. *Front Genet*. 2018;9:106.
86. Klass DM, Lauer N, Hay B, Kratzer W, Fuchs M. Arg 64 variant of the β3-adrenergic receptor is associated with gallstone formation. *Am J Gastroenterol*. 2007;102:2482.
87. Rho YH, Choi SJ, Lee YH, Ji JD, Song GG. The association between hyperuricemia and the Trp64Arg polymorphism of the beta-3 adrenergic receptor. *Rheumatol Int*. 2007;27:835.
88. Huang Q, Zhang LF, Cheng Y, et al. Trp64Arg (rs4994) polymorphism of β3-adrenergic receptor gene is associated with hyperuricemia in a Chinese male population. *Clin Chem Lab Med*. 2013;51:1755–1760.
89. Morcillo S, Rojo-Martínez G, Martín-Núñez GM, et al. Trp64Arg polymorphism of the ADRB3 gene predicts hyperuricemia risk in a population from southern Spain. *J Rheumatol*. 2010;37:417–421.
90. Qu HC, Zhang W, Liu YL, Wang P. Association between polymorphism of β3-adrenoceptor gene and overactive bladder: a meta-analysis. *Genet Mol Res*. 2015;14:2495–2501.
91. Ahles A, Engelhardt S. Polymorphic variants of adrenoceptors: pharmacology, physiology, and role in disease. *Pharmacol Rev*. 2014;66:598–637.
92. Kurokawa N, Young EH, Oka Y, et al. The ADRB3 Trp64Arg variant and BMI: a meta-analysis of 44 833 individuals. *Int J Obes*. 2008;32:1240–1249.
93. Hoffstedt J, Poirier O, Thorne A, et al. Polymorphism of the human β_3-adrenoceptor gene forms a well-conserved haplotype that is associated with moderate obesity and altered receptor function. *Diabetes*. 1999;48:203–204.
94. Teitsma CA, de la Rosette JJ, Michel MC. Are polymorphisms of the β3-adrenoceptor gene associated with an altered bladder function? *Neurourol Urodyn*. 2013;32:276–280.
95. Carlsson M, Orho-Melander M, Hedenbro J, Groop LC. Common variants in the β2-(Gln27Glu) and β3-(Trp64Arg)-adrenoceptor genes are associated with elevated serum NEFA concentrations and type II diabetes. *Diabetologia*. 2001;44:629–636.
96. Mentuccia D, Proietti-Pannunzi L, Tanner K, et al. Association between a novel variant of the human type 2 deiodinase gene Thr92Ala and insulin resistance: evidence of interaction with the Trp64Arg variant of the β-3-adrenergic receptor. *Diabetes*. 2002;51:880–883.
97. Corella D, Guillen M, Portoles O, et al. Gender specific associations of the Trp64Arg mutation in the β3-adrenergic receptor gene with obesity-related phenotypes in a Mediterranean population: interaction with a common lipoprotein lipase gene variation. *J Intern Med*. 2001;250:348–360.
98. Aoyama M, Shidoji Y, Saimei M, Tsunawake N, Ichinose M. Phenotypic linkage between single-nucleotide polymorphisms of β3-adrenergic receptor gene and NADH dehydrogenase subunit-2 gene, with special reference to eating behavior. *Biochem Biophys Res Commun*. 2003;309:261–265.
99. Halushka MK, Fan JB, Bentley K, et al. Patterns of single-nucleotide polymorphisms in candidate genes for blood-pressure homeostasis. *Nat Genet*. 1999;22:239–247.
100. Vrydag W, Alewijnse AE, Michel MC. Do gene polymorphisms alone or in combination affect the function of human β3-adrenoceptors? *Br J Pharmacol*. 2009;156:127–134.
101. Lehman DM, Hamlington J, Hunt KJ, et al. A novel missense mutation in ADRB3 increases risk for type 2 diabetes in a Mexican American family. *Diabetes Metab Res Rev*. 2006;22:331–336.

102. Voruganti VS, Cole SA, Ebbesson SO, et al. Genetic variation in APOJ, LPL, and TNFRSF10B affects plasma fatty acid distribution in Alaskan Eskimos. *Am J Clin Nutr.* 2010;91:1574–1583.
103. Hamilton G, Proitsi P, Jehu L, et al. Candidate gene association study of insulin signaling genes and Alzheimer's disease: evidence for SOS2, PCK1, and PPARγ as susceptibility loci. *Am J Med Genet B Neuropsychiatr Genet.* 2007;144:508–516.
104. Piétri-Rouxel F, St John Manning B, Gros J, Strosberg AD. The biochemical effect of the naturally occurring Trp64→ Arg mutation on human β3-adrenoceptor activity. *Eur J Biochem.* 1997;247:1174–1179.
105. Kimura K, Sasaki N, Asano A, et al. Mutated human β3-adrenergic receptor (Trp64Arg) lowers the response to β3-adrenergic agonists in transfected 3T3-L1 preadipocytes. *Horm Metab Res.* 2000;32:91–96.
106. Perfetti R, Hui H, Chamie K, et al. Pancreatic β-cells expressing the Arg64 variant of the β_3-adrenergic receptor exhibit abnormal insulin secretory activity. *J Mol Endocrinol.* 2001;27:133–144.
107. Candelore MR, Deng L, Tota LM, Kelly LJ, Cascieri MA, Strader CD. Pharmacological characterization of a recently described human β3-adrenergic receptor mutant. *Endocrinology.* 1996;137:2638–2641.
108. Isogaya M, Nagao T, Kurose H. Enhanced cAMP response of naturally occurring mutant of human β3-adrenergic receptor. *Jpn J Pharmacol.* 2002;88:314–318.
109. Li LS, Lönnqvist F, Luthman H, Arner P. Phenotypic characterization of the Trp64Arg polymorphism in the β3-adrenergic receptor gene in normal weight and obese subjects. *Diabetologia.* 1996;39:857–860.
110. Umekawa T, Yoshida T, Sakane N, Kogure A, Kondo M, Honjyo H. Trp64Arg mutation of β_3-adrenoceptor gene deteriorates lipolysis induced by β_3-adrenoceptor agonist in human omental adipocytes. *Diabetes.* 1999;48:117–120.
111. MacArthur MW, Thornton JM. Influence of proline residues on protein conformation. *J Mol Biol.* 1991;218:397–412.
112. Leineweber K, Büscher R, Bruck H, Brodde OE. β-Adrenoceptor polymorphisms. *Naunyn Schmiedebergs Arch Pharmacol.* 2004;369:1–22.

CHAPTER SIX

Mutations in chemokine receptors and AIDS

Jing Kou[a], **Yi-Qun Kuang**[b,*]
[a]International Education College, Henan University, Kaifeng, China
[b]Center for Translational Medicine, Huaihe Clinical College, Huaihe Hospital of Henan University, Kaifeng, China
*Corresponding author: e-mail address: yqkuang@vip.henu.edu.cn

Contents

Abstract

Chemokines are a class of chemotactic small molecule peptides whose receptors CCR5 and CXCR4 play important role in the entry of human immunodeficiency virus (HIV-1) into immune cells. Chemokines belong to G protein-coupled receptor superfamily containing seven hydrophobic transmembrane helices, causing physiological effects such as chemotaxis, immune regulation, antiviral immunity, regulation of hematopoiesis and angiogenesis, as well as cell growth and metabolism, through certain signaling pathways. Earlier studies have shown that HIV infects the human immune cells by binding to the CD4 receptor. Soon, it was discovered that HIV-1 enters into human immune cells by binding to another receptor, chemokine receptor, which acts as co-receptor for CD4 during the invasion of HIV-1 into cells. Since complex receptor binding is important for HIV-1 invasion, antagonizing the binding has become an attractive and rational drug design goal. Early studies sought to block the interaction between virus and the receptors by chemically modifying the CCR5 and CXCR4 ligands. Although drug treatment is widely used, drug treatment cannot cure AIDS; it can only inhibit the replication of the virus, and HIV/AIDS patients need to take drugs for life. In addition, anti-AIDS drugs also produce side effects such as diseases of the cardiovascular system, nervous system, and metabolic system. In 2006, the emergence of "Berlin patient" led researchers to focus on gene therapy in chemokine receptors. In 2006 and 2007, the attending physician of "Berlin patient" cured his AIDS by transplantation of the stem cells from a donor who was homozygous for the CCR5 Δ32 mutation. This review summarizes the research progress in the mutation in chemokine receptor of HIV/AIDS.

Progress in Molecular Biology and Translational Science, Volume 161
ISSN 1877-1173
https://doi.org/10.1016/bs.pmbts.2018.10.001

1. Introduction

The absorption of human immunodeficiency virus (HIV) is required for a receptor-mediated binding to the specific receptor of the host cell, CD4, on the surface of T4 lymphocytes, mononuclear macrophages, and dendritic cells and others. After the gp120 glycoprotein of HIV envelope binds to CD4 on the cell membrane, the conformation of gp120 changes to expose gp41, and the gp120-CD4 bimolecular complex binds to the C–C chemokine receptor type 5 (CCR5) or C–X–C motif chemokine receptor (CXCR4) on the target cell surface to form a CD4-gp120-CXCR4/CCR5 trimolecular complex. Disrupting the interaction between chemokine receptors and gp120 is an effective method for prevention of HIV infection.

In recent years, gene editing of CCR5/CXCR4 plays an important role in anti-HIV research.

In 2009, a clinical report from a German hospital aroused significant interest in the field. In this report, one Caucasian man diagnosed as acute myeloid leukemia (AML) and HIV-1 infection more than 10 years earlier and treated with HAART (highly active antiviral therapy) for 4 years was cured by stem-cell transplantation. The stem cell was collected from a donor who was homozygous for CCR5 Δ32 with 32-bp deletion.[1,2] Earlier studies have revealed that CCR5 Δ32 allele is deficient in expressing CCR5 protein at the plasma membrane, thus the interaction between the HIV and the host call is disrupted.[3] However, this therapeutic regimen was not working in another patient. A 27-year-old patient with HIV-1 infection and anaplastic large cell lymphoma was transplanted with stem cells from a donor who was also homozygous for the CCR5 Δ32 mutation. Three weeks after the transplantation without antiretroviral therapy, a rebound of 93,390 copies of HIV RNA per milliliter was detected.[4] These examples offered a possibility and idea to cure Acquired Immune Deficiency Syndrome (AIDS). From then on, gene editing of CCR5 is getting increased attention for anti-HIV research. Over the last few years, there are several methods for gene editing of CCR5. We will review three methods herein: engineered zinc finger nucleases (ZFNs), transcription activator-like (TAL) effector nucleases (TALENs) and clustered regularly interspaced short palindromic repeat (CRISPR)-Cas9 system, with the common goal to create CCR5-negative cells resistant to HIV-1 infection.

2. Disruption of CCR5 using engineered ZFNs

Each ZFN is a heterodimer with each subunit comprised of two domains: DNA-binding domain and DNA-cleaving domain. The DNA-binding domain contains a chain of two zinc finger modules, which can recognize a unique hexamer sequence of DNA. Two-finger modules are dimerized to form a zinc finger protein (ZFP) with a high specificity of $\geq$24 bp. The DNA-cleaving domain comprises of a FokI endonuclease. Two FokI must dimerize together to cleavage the location recognized by ZFP at a high level of specificity. Once the double-strand breaks (DSBs) in target genes has been achieved, the cell's endogenous homologous directed recombination (HDR) machinery and nonhomologous end joining (NHEJ) will then be stimulated. In NHEJ, cells essentially smooth the ends of the broken DNA and then pull them closer together, which often results in a frameshift. By HDR, the cell attempts to repair the break using the corresponding DNA sequence on the other strand as a template. By providing their own templates, the desired sequence can be inserted inadvertently in this gap. As such, a pair of ZFNs must be designed for each side of the cut location on opposite strands. DSBs can be precisely generated in the targeted genomic region, only when dual binding of ZFNs occurs does dimerization and cleavage ensue.[5–7]

Using engineered ZFNs, many scientists managed to create *CCR5* gene disruption to seek the possibility of HIV/AIDS therapy. The ZFNs recognized the CCR5 coding region and generated a DSB at a target site. The permanent disruption of *CCR5* gene resulted in interfering HIV infection in primary human $CD4^+$ T lymphocytes.

Perez et al. have established two approaches both *in vitro* and in an *in vivo* mouse model using CCR5 ZFNs against HIV-1 infection in $CD4^+$ T cells. According to Urnov et al.'s method,[7] they designed a large series of ZFNs targeted genome encoding the first transmembrane domain (TM1 spanning residues Arg31 to Asn57) of CCR5, upstream of the natural CCR5 Δ32 mutation. This study showed that transient expression of CCR5 ZFNs using adenoviral vector can cause a mean frequency of 50% of *CCR5* alleles with permanent and specific disruption in primary human $CD4^+$ T cell culture, thus the human $CD4^+$ T cells with engineered CCR5 ZFNs possess a long-term and stable resistance to HIV-1 infection *in vitro*. The feasibility of this method was demonstrated in a NOD/Shi-*scid*/IL-2Rγ^{null} (NOG) mouse with HIV-1 infection. HIV-infected mice engrafted with CCR5 ZFN-modified $CD4^+$ T cells have lower viremia and higher peripheral $CD4^+$ T cell counts. This study showed an efficient gene editing of HIV-1 chemokine receptors using ZFNs in an animal model and provided a possibility of gene therapy for HIV-1 patients in clinical trials. Transfer of *ex vivo* expanded ZFN-modified autologous $CD4^+$ T cells in HIV patients may be an attractive approach for the treatment of HIV-1 infection.[8]

On the other hand, the attempt to edit *CCR5* gene using ZFNs is not limited in $CD4^+$ T cell culture or an animal model. Li et al. demonstrated *CCR5* gene disruption in $CD34^+$ hematopoietic stem/progenitor cells (HSPCs) isolated from granulocyte colony-stimulating factor-mobilized adult peripheral blood (delivery using adenoviral vector) and cord blood (via nucleofection of plasmid DNA) less than 5% and 17% of total alleles within the cell population, respectively. To improve the activity of ZFNs in HSPC and the efficiency of CCR5 disruption, they subsequently treated the cells with protein kinase C (PKC) activators and demonstrated enhancement of *CCR5* gene disruption.[9]

Along with the improved ZFN-mediated CCR5 disruption technology, Sangamo BioSciences, Inc. released the data from HIV clinical trial at CROI (Conference on Retroviruses and Opportunistic Infections) 2013. The company optimized the Fok1 enzyme of ZFNs and modified an autologous $CD4^+$ T cell product named SB-728-T, in which the genome editing of CCR5 was mediated via ZFNs. SB-728-T with a permanently disrupted CCR5 are resistant to infection by the most common strain of HIV. In the clinical trial, the safety and persistence of SB-728-T in the blood were evaluated. Both acute and long-term increase in total $CD4^+$ T cell counts were recorded from HIV-infected subjects treated with SB-728-T. The viral load is reduced to undetectable levels. The level of $CD4^+$ T cell reconstitution is significantly enhanced. This clinical trial provides a powerful mean for the treatment of human monogenic hereditary diseases and achieved targeted modification of the genome. Even to this day, Sangamo BioSciences, Inc. is committed to build a new therapeutic platform using ZFN technology to regulate target gene for other diseases, such as Hemophilia B, MPS I (Hurler syndrome), and MPS II (Hunt's syndrome).

3. Disruption of CCR5 using engineered TALENs

TALEN is another flexible and an efficient restriction enzyme for achieving gene editing following ZFN, and structurally similar to ZFN.

A TALEN recognizes specific DNA base pairs by means of the TAL effector (TALE), a natural protein secreted by plant pathogenic bacteria *Xanthomonas* sp., including the N-terminal translocation signal, the C-terminal nuclear localization signal, the activation domain, and the intermediate tandem repeat region, which is composed of a plurality of repeating sequence units arranged in series. Each repeat unit generally contains 33–35 highly conserved amino acids, wherein 12th and 13th amino acids are variable and are referred to as repeat-variable diresidues (RVD).[10,11] RVD mediates specific nucleotide recognition of TALE that a two amino acids repeat specifically recognizes and binds one base pair. The following recognition preferences have been experimentally identified: HD = C; NG = T; NI = A; NS = A, C, G, or T; NN = A or G; and IG = T.[12,13] Based on these recognition preferences, TALE proteins with specific recognition DNA sequences can be customized by multiple TALE repeat units in series to target DNA sequences of interest.[14] A nonspecific endonuclease FokI is constructed at the C-terminus of TALE protein to constitute a TALEN monomer used for targeted genome editing. FokI nuclease is responsible for cleavage of the target DNA; thereby FokI dimerization from two TALENs causes DNA DSBs and inducing DNA damage of the cells. Repair mechanisms enable editing of genomic target sites.[15] Compared to ZFN, TALEN can target longer gene sequences and is easier to construct.

In Ru et al.'s study, they constructed a cell-penetrating peptide (CPP)-based system to deliver the engineered ZFNs and TALENs into target cell. CPPs have been widely used to introduce various biomolecules, including plasmid DNA, oligonucleotides, short interfering RNA, peptide nucleic acid, proteins and peptides, as well as liposome nanoparticles, into all cell types both *in vivo* and *in vitro* with high transduction efficiency and controlled administration.[16] Among the CPPs, TAT is a short peptide consisting of 11 amino acids (YGRKKRRQRRR) derived from the protein transactivator of transcription (TAT) domain of HIV-1 used for cell penetrating.[17] Functional TAT-ZFN and TAT-TALEN proteins were generated by fusing the cell-penetrating TAT peptide to ZFN and TALEN, respectively, which are able to penetrate cells and directly cut the *CCR5* gene for preventing HIV infection. They demonstrated that *CCR5* gene disruption frequencies of up to 16% and 5% in HeLa cells and in human-induced pluripotent stem cells, respectively, whereas standard TALENs show no effects on *CCR5* disruption under same conditions.[18]

Mussolino et al. designed engineered ZFNs and TALENs targeting human CCR5 locus. The efficiency of CCR5 disruption was analyzed in

HEK293T cells and primary newborn human foreskin fibroblasts with average frequencies up to 24% and 16%, respectively.[19]

A total of 28 novel CCR5-TALENs targeting multiple regions of *CCR5* gene were designed and expressed by Shi et al.[20] The efficiency of *CCR5* gene disruption was analyzed in the GHOST-CCR5-CXCR4 cells and human primary $CD4^+$ T cells. They have identified one (CCR5-TALEN-515) with the highest efficiency. Furthermore, they also demonstrated that CCR5-TALENs with higher nuclease activity, specificity, and lower cytotoxicity, as well as less off-target activity compared to CCR5-ZFN.[20]

To further confirm the disruption in CCR5 expression, Tang et al. established a new animal model for CCR5 disruption. Wild rabbits can be infected by HIV-1 persistently. A rabbit cell line coexpressing human CD4 and CCR5 may be an ideal animal model for HIV/AIDS study. Therefore, they constructed TALENs targeting rabbit CCR5 gene and a homologous targeting vector. TALEN mRNAs and donor DNA were then coinjected into fertilized oocytes. After 3–5 days, 24 embryos were collected and analyzed for *CCR5* mutation. All 24 embryos were detected with CCR5 knockouts and 5 were human CD4 and CCR5 knock-ins.

The news about the clinical trial directly utilized by TALENs are not found. But T cells were edited using the TALEN technique to target the CD123 antigen. Such antigens are highly expressed in AML cells and blast-like dendritic cell tumor (BPDCN) cells. Both diseases often develop in the bone marrow and can threaten the lives of patients in the short term. This is the first universal CAR-T therapy to enter U.S. clinical trials. Cellectis announced that its universal CAR-T therapy, UCART123, has been approved by the US FDA for clinical trials. This is also the first such product to be approved by the US FDA for clinical trials. In 2015, scientists successfully implemented the world's first treatment for infant leukemia using this gene-editing technology. After treatment, the test results showed that the cancer cells in the patient have been completely eliminated. Unfortunately, on September 5, 2017, the FDA urgently suspended the ongoing CAR-T trial of Cellectis for safety reasons. The company's first human test was a 78-year-old patient with acute plasmacytoid dendritic cell tumors. The subject died 8 days after the first injection, and the cause of death was cytokine release syndrome and capillary leak syndrome.

ZFNs and TALENs technology can cause serious consequences such as off-target, cytotoxicity, etc., which can lead to cancerous cells.

4. Disruption of CCR5 using engineered CRISPR-Cas9

The CRISPR-Cas system is one of the most exciting in all gene-editing technologies. CRISPR is an adaptive immune defense formed by bacteria and archaebacteria during its long-term evolution. It can be used against invading viruses and foreign DNA, which is widely found in most bacteria and almost all archaebacteria. The CRISPR mechanism recognizes foreign DNA by complementary pairing of bases with RNA, and then precisely cleaves the DNA double strand of the virus or phage with the help of the CRISPR-associated (Cas) endonuclease to break the DNA double strand and prevent the replication and synthesis of virus or phage that protect the bacteria themselves.

The CRISPR-Cas system is mainly composed of the Cas protein gene, the leader sequence, and the CRISPR locus. CRISPR loci typically consist of several noncontiguous direct repeats separated by stretches of variable sequences called spacers (which mostly correspond to segments of captured viral and plasmid sequences) and are often adjacent to *cas* genes. Repeats are arranged in cluster, most of them palindromic to form stable RNA stem-loop secondary structure. The leader sequence is located 5′ to most CRISPR loci and has been suggested to act as a recognition sequence for the addition of new spacers or a promoter to initiate transcription of the CRISPR locus. The *cas* gene encodes the endonuclease Cas protein functioned together with guide RNA to locate and cleave DNA of phage and plasmids at conserved sequences called proto-spacer adjacent motifs. Among several Cas protein types, Cas9 is applied really often to reengineer into a more manageable system to site specifically modify gene or genome of interest.[21–24] To form a functional complex for site-specific cleavage, Cas9 needs two additional RNA molecules: CRISPR RNA (crRNA) and *trans*-acting CRISPR RNA (tracrRNA). The crRNA binds to tracrRNA by base pairing to form a tracrRNA/crRNA complex that directs the nuclease Cas9 protein to introduce DSBs at a target sequence site paired with the crRNA. However, recent studies have found that these two RNAs can be "modified" into a single-guide RNA (sgRNA). This sgRNA is sufficient to help the Cas9 endonuclease cleave the DNA.[25] Based on recent studies, in many types of cells and organisms, this RNA-mediated Cas9 digestion can function normally and complete the cleavage reaction at specific sites on the complete genome. This makes it easy to carry out subsequent genomic modifications. Cells usually repair DNA that undergoes DSBs in two ways: homologous recombination and NHEJ, but during the repair process, the cells may modify the repair site or insert new genetic information.

In 2013, two laboratories used this technology to successfully target the endogenous genes of human and mouse cells.[26,27] Hu et al. previously used CRISPR to remove HIV from the human genome.[28]

Researches also used this technology to modify CCR5 gene. Xu et al. have established a system based on CRISPR/Cas9 technology in human fetal liver HSPCs. In the study, they successfully achieved gene editing using CRISPR/Cas9 and nonviral transfection system in HSPCs with high cleavage efficiency and low off-target effect. CCR5 disruption occurred with an average of 42% and 27% cleavage efficiency in K562 cells and in human $CD34^+$ HSPCs *in vitro*, respectively. They also confirmed that these modified HSPC cells are able to block HIV infection after transplantation into long-term reconstituted NOD/*Prkdc*^{scid}/*IL-2Rγ*^{null} (NPG) mice for more than 1 year and observed a significant antiviral effect *in vivo*.[29]

Ye et al. established induced pluripotent stem cells (iPSCs) of CCR5 Δ32 homozygous using TALENs or CRISPR-Cas9 together with piggyBac technology. TALENs or CRISPR-Cas9-mediated double-strand DNA breaks results in up to 100% targeting of wild-type iPSCs on one allele of which biallelic targeting occurs at an average of 14% with TALENs and 33% with CRISPR. Using piggyBac transposon technology to generate CCR5 Δ32 mutation results in seamless modification without detectable exogenous sequences. These modified iPSCs were differentiated into monocytes and macrophages. These cells remained their resistance to HIV-1 infection.[30]

In Li et al.'s study, a CRISPR-Cas9 system delivered by adenovirus was first constructed for inducing CCR5 disruption in primary $CD4^+$ T lymphocytes, and the positively transduced cells were resistant to HIV-1.[31]

Compared to ZFNs and TALENs, one of the advantages of CRISPR system is its high cell transfection efficiency. Moreover, CRISPR system can reduce the residual sequences during gene editing. However, using these three technologies to modify *CCR5* gene may not be able to cure HIV/AIDS, because targeting CCR5 will not be resistant to CXCR4-tropic (X4) HIV isolates but only to CCR5-tropic (R5) isolates. CXCR4-tropic (X4) HIV isolate uses the CXCR4 instead of CCR5 to invade immune cells. Thus, mutation in *CXCR4* gene using these three technologies is clearly an alternative strategy for HIV/AIDS therapy that is worth investigating.

5. Disruption of CXCR4

Wilen et al. engineered ZFNs into human T cells including CCR5 Δ32 $CD4^+$ T cells that resulted in efficient CXCR4 disruption. These modified T cells can proliferate normally and are resistant to X4-tropic HIV-1 infection. They also demonstrated that CXCR4 disruption provides a short-term protection from X4-tropic HIV-1 in a humanized mouse model.[32] Yuan et al. generated ZFNs for CXCR4 disruption in $CD4^+$ T cells with stable HIV-1 resistance. The modified $CD4^+$ T cells can be enriched during HIV-1 infection in tissue culture and in peripheral blood mononuclear cell transplanted NSG mice with lower viral levels than in unmodified $CD4^+$ T cells.[33] Hou et al. generated CRISPR/Cas9-mediated genome-editing system for CXCR4 disruption in human primary $CD4^+$ T cells, and the cells were modified to HIV-1 resistance with high specificity and low off-target effects.[34]

Mutation in *CXCR4* also provides an alternative way for preventing HIV-1 infection remaining high precision and efficiency, as well as long term. As mutation in CCR5, CXCR4 disruption will not be resistant to CCR5-tropic (R5) HIV isolates.

6. Disruption of both CCR5 and CXCR4

Binding to CD4 combined with CCR5 or CXCR4 is a necessary condition for HIV entry into host cell. CCR5 or CXCR4 disruption can be only resistant to R5-tropic or X4-tropic HIV strains, respectively.[35,36] Simultaneous disruption of CCR5 and CXCR4 in T cells can result in blocking HIV-1 infection both with R5-tropic strain and X4-tropic strain. Didigu et al. modified *CCR5* and *CXCR4* genes using ZFNs in primary human $CD4^+$ T cells.[37] These modified cells can still proliferate and are resistant to both R5- and X4-tropic HIV-1 strain *in vitro*. Moreover, these cells were introduced into a humanized mouse model of HIV-1 infection and maintain resistance of R5- and X4-tropic HIV-1 infection.[37]

Liu et al. constructed CRISPR-Cas9 system targeting both CXCR4 and CCR5 delivered by lentivirus in TZM-bl cells derived from epithelial origin, Jurkat T cells, and human $CD4^+$ T cells to examine the knockout efficiency. They demonstrated that the modified CRISPR-Cas9 system can successfully induce gene editing in all three cell types and protect the cells from CXCR4-tropic or CCR5-tropic HIV-1 infections. They also demonstrated that these modified cells do not show off-target editing and obvious cytotoxic effects in human primary $CD4^+$ T cells with simultaneous disruption of *CXCR4* and *CCR5*.[38]

In summary, simultaneous disruption of CXCR4 and CCR5 using genome-editing technologies may provide an effective approach for the cure of HIV-1 infection with long-term effect and no cytotoxicity.

7. Conclusions

Chemokine receptor-mediated signaling pathways are involved in HIV-1 infection. The in-depth study of chemokine receptors will serve as the theoretical and experimental basis for the treatment of HIV/AIDS. Chemokine receptors are relatively stable cell-targeting factors in HIV-1 infection. Although a large number of chemokine receptor antagonists are developed to target the different stages of HIV-1 invasion, research progress is still slow. So far, the main approach is combination antiretroviral therapy, which is effective in the long term only when multiple antiretroviral agents are used. Therefore, chemokine receptor targeting strategies can be successful only if they are combined with traditional antiretroviral therapies rather than being used alone. The mutation in coreceptors provides a new direction for the treatment of infection. HIV-1 gene therapy using ZFN, TALEN, and CRISPR-Cas9 has been confirmed to successfully achieve CCR5 or CXCR4 disruption in human cells *in vitro* and humanized mouse model *in vivo*. The modified cells remain low viral load and proliferation without cytotoxicity. Furthermore, simultaneous disruption of CXCR4 and CCR5 has been similarly confirmed in human cell lines and humanized mouse model with low viral load and negligible cytotoxicity. These suggest that mutations in chemokine receptors CCR5 and CXCR4 for HIV-1 infection provide an effective alternative to drugs for the long-term treatment of HIV/AIDS.

References

1. Hutter G, Nowak D, Mossner M, et al. Long-term control of HIV by *CCR5* Delta32/Delta32 stem-cell transplantation. *N Engl J Med*. 2009;360:692–698.
2. Allers K, Hutter G, Hofmann J, et al. Evidence for the cure of HIV infection by CCR5 Δ32/Δ 32 stem cell transplantation. *Blood*. 2011;117:2791–2799.
3. Dean M, Carrington M, Winkler C, et al. Genetic restriction of HIV-1 infection and progression to AIDS by a deletion allele of the CKR5 structural gene. Hemophilia Growth and Development Study, Multicenter AIDS Cohort Study, Multicenter Hemophilia Cohort Study, San Francisco City Cohort, ALIVE Study. *Science*. 1996;273:1856–1862.

4. Kordelas L, Verheyen J, Beelen DW, et al. Shift of HIV tropism in stem-cell transplantation with *CCR5* Delta32 mutation. *N Engl J Med.* 2014;371:880–882.
5. Durai S, Mani M, Kandavelou K, Wu J, Porteus MH, Chandrasegaran S. Zinc finger nucleases: custom-designed molecular scissors for genome engineering of plant and mammalian cells. *Nucleic Acids Res.* 2005;33:5978–5990.
6. Porteus MH, Carroll D. Gene targeting using zinc finger nucleases. *Nat Biotechnol.* 2005;23:967–973.
7. Urnov FD, Miller JC, Lee YL, et al. Highly efficient endogenous human gene correction using designed zinc-finger nucleases. *Nature.* 2005;435:646–651.
8. Perez EE, Wang J, Miller JC, et al. Establishment of HIV-1 resistance in CD4+ T cells by genome editing using zinc-finger nucleases. *Nat Biotechnol.* 2008;26:808–816.
9. Li L, Krymskaya L, Wang J, et al. Genomic editing of the HIV-1 coreceptor CCR5 in adult hematopoietic stem and progenitor cells using zinc finger nucleases. *Mol Ther.* 2013;21:1259–1269.
10. Boch J, Bonas U. Xanthomonas AvrBs3 family-type III effectors: discovery and function. *Annu Rev Phytopathol.* 2010;48:419–436.
11. Boch J. TALEs of genome targeting. *Nat Biotechnol.* 2011;29:135–136.
12. Boch J, Scholze H, Schornack S, et al. Breaking the code of DNA binding specificity of TAL-type III effectors. *Science.* 2009;326:1509–1512.
13. Moscou MJ, Bogdanove AJ. A simple cipher governs DNA recognition by TAL effectors. *Science.* 2009;326:1501.
14. Zhang F, Cong L, Lodato S, Kosuri S, Church GM, Arlotta P. Efficient construction of sequence-specific TAL effectors for modulating mammalian transcription. *Nat Biotechnol.* 2011;29:149–153.
15. Zhang M, Wang F, Li S, Wang Y, Bai Y, Xu X. TALE: a tale of genome editing. *Prog Biophys Mol Biol.* 2014;114:25–32.
16. Morris MC, Deshayes S, Heitz F, Divita G. Cell-penetrating peptides: from molecular mechanisms to therapeutics. *Biol Cell.* 2008;100:201–217.
17. Vyas PM, Payne RM. TAT opens the door. *Mol Ther.* 2008;16:647–648.
18. Ru R, Yao Y, Yu S, et al. Targeted genome engineering in human induced pluripotent stem cells by penetrating TALENs. *Cell Regen (Lond).* 2013;2–5.
19. Mussolino C, Alzubi J, Fine EJ, et al. TALENs facilitate targeted genome editing in human cells with high specificity and low cytotoxicity. *Nucleic Acids Res.* 2014;42:6762–6773.
20. Shi B, Li J, Shi X, et al. TALEN-mediated knockout of CCR5 confers protection against infection of human immunodeficiency virus. *J Acquir Immune Defic Syndr.* 2017;74:229–241.
21. Garneau JE, Dupuis ME, Villion M, et al. The CRISPR/Cas bacterial immune system cleaves bacteriophage and plasmid DNA. *Nature.* 2010;468:67–71.
22. Horvath P, Barrangou R. CRISPR/Cas, the immune system of bacteria and archaea. *Science.* 2010;327:167–170.
23. Sorek R, Kunin V, Hugenholtz P. CRISPR—a widespread system that provides acquired resistance against phages in bacteria and archaea. *Nat Rev Microbiol.* 2008;6:181–186.
24. Charpentier E, Doudna JA. Biotechnology: rewriting a genome. *Nature.* 2013;495:50–51.
25. Jinek M, Chylinski K, Fonfara I, Hauer M, Doudna JA, Charpentier E. A programmable dual-RNA-guided DNA endonuclease in adaptive bacterial immunity. *Science.* 2012;337:816–821.
26. Cong L, Ran FA, Cox D, et al. Multiplex genome engineering using CRISPR/Cas systems. *Science.* 2013;339:819–823.
27. Mali P, Yang L, Esvelt KM, et al. RNA-guided human genome engineering via Cas9. *Science.* 2013;339:823–826.

28. Hu W, Kaminski R, Yang F, et al. RNA-directed gene editing specifically eradicates latent and prevents new HIV-1 infection. *Proc Natl Acad Sci U S A*. 2014;111: 11461–11466.
29. Xu L, Yang H, Gao Y, et al. CRISPR/Cas9-mediated CCR5 ablation in human hematopoietic stem/progenitor cells confers HIV-1 resistance in vivo. *Mol Ther*. 2017; 25:1782–1789.
30. Ye L, Wang J, Beyer AI, et al. Seamless modification of wild-type induced pluripotent stem cells to the natural CCR5 Δ32 mutation confers resistance to HIV infection. *Proc Natl Acad Sci U S A*. 2014;111:9591–9596.
31. Li C, Guan X, Du T, et al. Inhibition of HIV-1 infection of primary CD4+ T-cells by gene editing of CCR5 using adenovirus-delivered CRISPR/Cas9. *J Gen Virol*. 2015; 96:2381–2393.
32. Wilen CB, Wang J, Tilton JC, et al. Engineering HIV-resistant human CD4+ T cells with CXCR4-specific zinc-finger nucleases. *PLoS Pathog*. 2011;7, e1002020.
33. Yuan J, Wang J, Crain K, et al. Zinc-finger nuclease editing of human cxcr4 promotes HIV-1 $CD4^+$ T cell resistance and enrichment. *Mol Ther*. 2012;20:849–859.
34. Hou P, Chen S, Wang S, et al. Genome editing of CXCR4 by CRISPR/cas9 confers cells resistant to HIV-1 infection. *Sci Rep*. 2015;5, 15577.
35. Alkhatib G, Combadiere C, Broder CC, et al. CC CKR5: a RANTES, MIP-1α, MIP-1β receptor as a fusion cofactor for macrophage-tropic HIV-1. *Science*. 1996;272: 1955–1958.
36. Bleul CC, Farzan M, Choe H, et al. The lymphocyte chemoattractant SDF-1 is a ligand for LESTR/fusin and blocks HIV-1 entry. *Nature*. 1996;382:829–833.
37. Didigu CA, Wilen CB, Wang J, et al. Simultaneous zinc-finger nuclease editing of the HIV coreceptors *ccr5* and *cxcr4* protects $CD4^+$ T cells from HIV-1 infection. *Blood*. 2014;123:61–69.
38. Liu Z, Chen S, Jin X, et al. Genome editing of the HIV co-receptors CCR5 and CXCR4 by CRISPR-Cas9 protects $CD4^+$ T cells from HIV-1 infection. *Cell Biosci*. 2017;7:47.

CHAPTER SEVEN

Physiology, pharmacology, and pathophysiology of neuropeptide S receptor

Zheng-Rui Zhang[a,b], Ya-Xiong Tao[a,c,*]
[a]Department of Anatomy, Physiology and Pharmacology, College of Veterinary Medicine, Auburn University, Auburn, AL, United States
[b]Key Laboratory of Marine Genetics and Breeding, Ministry of Education, College of Marine Life Sciences, Ocean University of China, Qingdao, China
[c]Center for Neuroscience Initiative, Auburn University, Auburn, AL, United States
*Corresponding author: e-mail address: taoyaxi@auburn.edu

Contents

Abstract

Neuropeptide S receptor 1 (NPSR1), originally named G protein-coupled receptor 154 (GPR154), was deorphanized in 2002 with neuropeptide S identified as the endogenous ligand. NPSR1 is primarily expressed in bronchus, brain as well as immune cells.

Progress in Molecular Biology and Translational Science, Volume 161
ISSN 1877-1173
https://doi.org/10.1016/bs.pmbts.2018.11.001

It regulates multiple physiological processes, including immunoregulation, locomotor activity, anxiety, arousal, learning and memory, and food intake and energy balance. SNPs of *NPSR1* are significantly associated with several diseases, including asthma, anxiolytic and arousal disorders, and rheumatoid arthritis. This chapter will summarize studies on NPSR1, including its molecular structure, tissue distribution, physiology, pharmacology, and pathophysiology.

Abbreviations

cAMP cyclic adenosine monophosphate
CHO Chinese hamster ovary cells
EC_{50} concentration that results in half maximal stimulation
ERK1/2 extracellular signal-regulated kinases 1/2
GPR154 G protein-coupled receptor 154
h human
HEK293 human embryonic kidney 293 cells
IBD inflammatory bowel disease
IgE immunoglobulin E
IP3 inositol 1,4,5-trisphosphate
MMP10 matrix metallopeptidase 10
NPS neuropeptide S
NPSR1 neuropeptide S receptor 1
REM rapid eye movement
SNP single-nucleotide polymorphism
TIMP3 tissue inhibitor of metallopeptidase inhibitor 3
WT wild type

1. Introduction

Asthma is a serious noncommunicable disease, which can lead to inflammation of airways of the lungs. Worldwide, about 358 million people suffered from chronic asthma in 2015.[1] According to data from Centers for Disease Control and Prevention of the United States, about 24.6 million people, including 6.1 million children, suffer from asthma in 2016. The annual economic cost of asthma in 2007 is about 56 billion dollars in the United States alone (https://www.cdc.gov/vitalsigns/asthma/, accessed October 20, 2018).

Asthma is usually caused by both environmental and genetic factors.[2] Neuropeptide S (NPS) and its receptor, neuropeptide S receptor 1 (NPSR1), were identified as genetic factors predisposing to asthma. NPS, a neuropeptide of 20 amino acids, with the primary sequence of

SFRNGVGTGMKKTSFQRAKS (in human), was first discovered in 2002.[3] The sequences of NPS peptide precursor are highly conserved among all terrestrial vertebrates, from amphibians to mammals,[4] and NPS is named after the highly conserved serine at the N-terminus.[5] So far, there is no report of any other endogenous ligand for NPSR1.

NPS plays important roles in the central nervous system. Studies show that NPS promotes arousal and inhibits all stages of sleep,[5,6] while it is also involved in reducing anxiety.[5–9] In addition, NPS can control food intake to maintain energy balance in rodents[10–12] and chicks.[13] Moreover, facilitation of spatial memory and learning may also be one of the functions of NPS.[12,14] Mutations in *NPS* can also modify asthma risk.[15].

G protein-coupled receptor 154 (GPR154), originally identified as an orphan G protein-coupled receptor (GPCR), also called GPRA (G protein-coupled receptor for asthma),[16] VRR1 (vasopressin receptor-related receptor 1),[17] was deorphanized in 2002[3] with NPS identified as the endogenous ligand. It has been named as NPSR1 by Human Gene Nomenclature Committee.

NPSR1 is localized on human chromosome 7p14–15 related to asthma.[16] *NPSR1* single-nucleotide polymorphisms (SNPs) were found to be closely associated with asthma and high serum immunoglobulin E (IgE) levels in several populations.[18–25] In addition, *NPSR1* SNPs are also related to several other diseases such as rheumatoid arthritis and inflammatory bowel disease (IBD).[26–28] Different combinations of *NPS* and *NPSR1* SNPs lead to various results of downstream signaling and genetic risk of disease including asthma.[15]

In this chapter, we will summarize the studies on NPSR1, including its molecular cloning and tissue distribution, physiology, pharmacology, and pathophysiology.

2. Molecular cloning and localization of NPSR1

2.1 Molecular cloning of NPSR1

In order to identify susceptibility locus for asthma-related traits on chromosome 7p, Laitinen et al. found two genes by using a hierarchical genotyping design.[16] One of them is *NPSR1* that has two transcripts, NPSR1-A and NPSR1-B. One year later, five additional shorter transcripts of NPSR1, named NPSR1-Bshort, NPSR1-C, NPSR1-D, NPSR1-E, NPSR1-F, were identified.[29] Of all the isoforms coded by transcripts, only the two full-length variants, NPSR1-A and NPSR1-B, are transported to the plasma membrane, while all the other shorter isoforms are retained in intracellular compartments,[29] consistent with findings in other GPCRs that intracellular retention is common in both wild-type (WT) and mutant GPCRs.[30,31] These two main transcripts of *NPSR1* contain nine exons with alternative 3′ exons, *NPSR1-A* encoding a protein of 371 amino acids while *NPSR1-B* encoding a protein of 377 amino acids (Fig. 1).

The DNA sequences encoding NPS peptide precursor are conserved among all terrestrial vertebrates but absent from fish.[4] However, NPSR1 or NPSR1-like sequences are found not only in terrestrial vertebrates but also in acorn worm and lancelet.[32] The amino acid sequences of vertebrate NPSR1 are shown in Fig. 2A. The only difference between isoform A and B of human (h) NPSR1 is at the C-termini (see also Fig. 1), with 93% identity. The homologies of NPSR1 among different species are summarized in Fig. 2B. Mouse and rat NPSR1 share 89% identity with hNPSR1-A. Monkey NPSR1 shares 97% identity with hNPSR1-A. Phylogenetic analysis indicates that hNPSR1 is highly homologous with those of other mammalian species at the amino acid level (Fig. 2C).

2.2 Tissue distribution of NPSR1

The distribution of NPSR1 expression in human tissues has been studied and shown to be ubiquitously expressed. Northern blotting showed that all eight tissues studied, including spleen, thymus, prostate, testis, uterus, small intestine, colon, lung, as well as peripheral leukocytes, express NPSR1.[16] Using purified antibodies raised again the divergent C-termini of the two isoforms, immunohistochemistry showed that isoform A is predominantly expressed in smooth muscle cells, whereas isoform B is predominantly expressed in apical epithelial cells of bronchus, gut, and skin.[16,29]

Vendelin et al. investigated the expression of NPSR1 at both mRNA and protein levels in more detail.[29] In paraffin sections of 30 human tissues, in situ hybridization shows positive signals for *NPSR1* in epithelial cells of bronchus, the gastrointestinal tract (esophagus, stomach, small intestine, colon), and skin. Positive signals also exist in submucosal epithelial cells of spleen, kidney, pancreas, prostate, uterus, breast, heart muscle, and in some glandular epithelia.[29]

In Western blot analyses, NPSR1-A is expressed in all tissues with smooth muscle include uterine muscle, colon muscle, and prostate, whereas NPSR1-B is distributed in all tissues studied, except for skeletal muscle.[29]

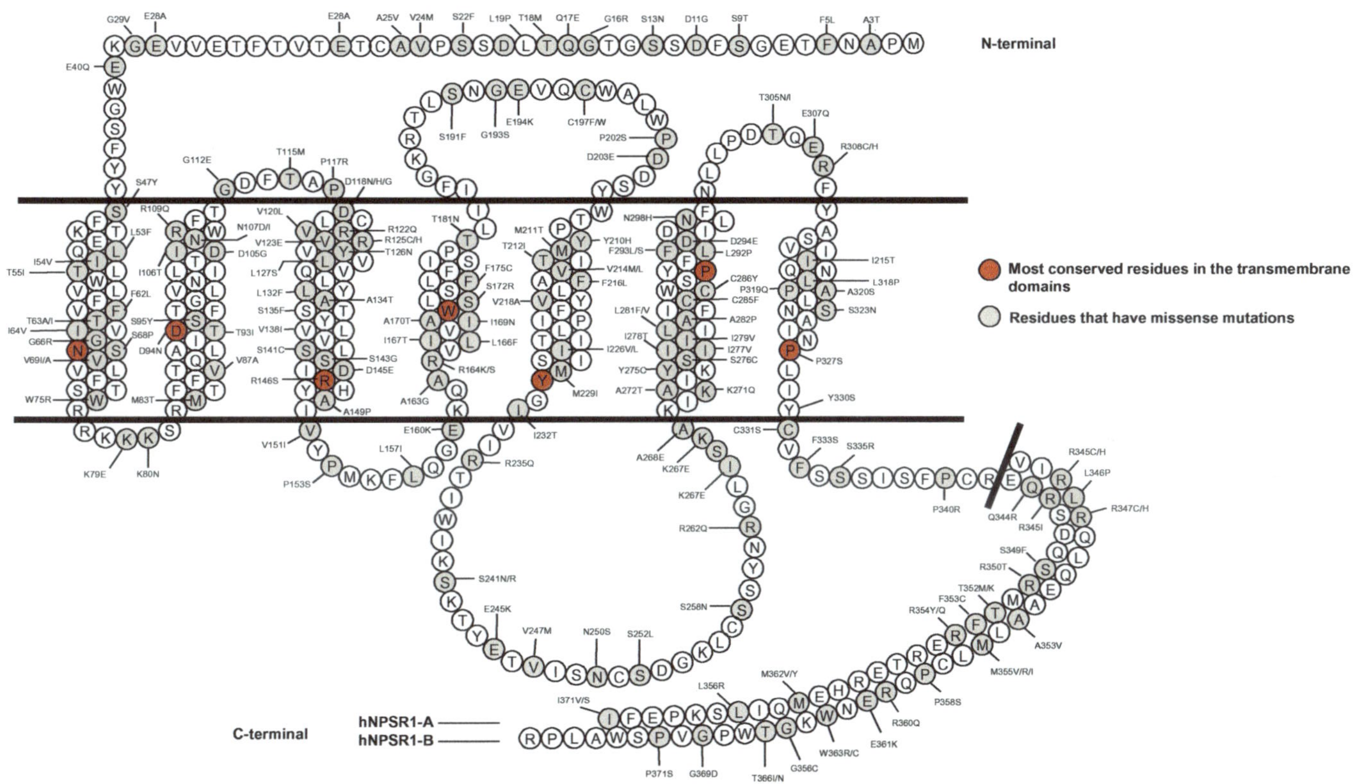

Fig. 1 Two-dimensional illustration of human NPSR1-A and NPSR1-B. The most conserved residues in the transmembrane domains are shaded in *red*. Residues that have naturally occurring mutations or polymorphisms identified by Exome Aggregation Consortium database are shaded in *gray*.

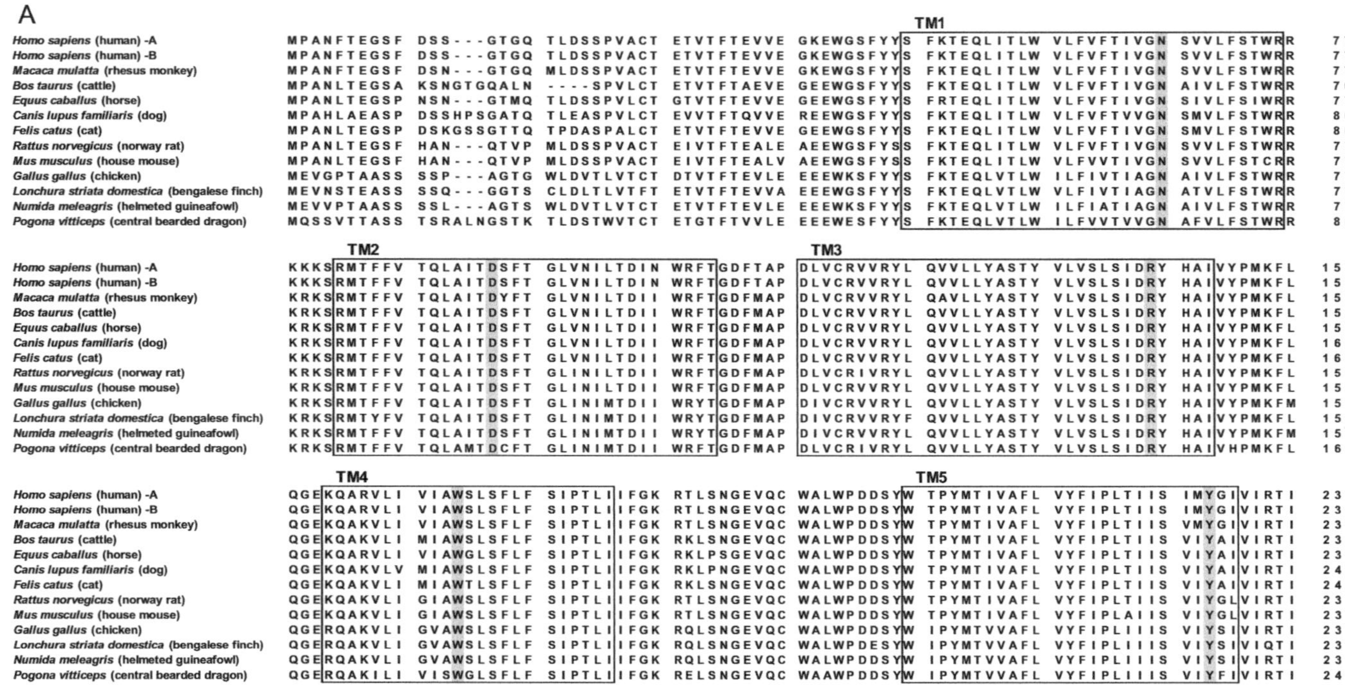

Fig. 2 Sequence analysis of NPSR1 in different species. (A) Alignment of the amino acid sequences between hNPSR1-A and B and NPSR1 cloned or predicted in other species. The transmembrane domains are *boxed*. The most conserved residues in the transmembrane domains are shaded in *gray*. (B) Homology and (C) phylogenetic analysis of NPSR1 amino acid sequences from different species. NCBI accession numbers for the proteins are as follows: human-A (*Homo sapiens*), NP_997055.1; human-B (*Homo sapiens*), NP_997056.1; monkey (*Macaca mulatta*), NP_001028114.1; cattle (*Bos taurus*), NP_001179906.1; horse (*Equus caballus*), XP_001501106.1; dog (*Canis lupus familiaris*), XP_022283280.1; cat (*Felis catus*), XP_011278803.1; mouse (*Mus musculus*), NP_783609.1; rat (*Rattus norvegicus*), NP_001100278.1; chicken (*Gallus gallus*), AQZ59723.1; guineafowl (*Numida meleagris*), XP_021244348.1; finch (*Lonchura striata domestica*), XP_021408948.1; bearded dragon (*Pogona vitticeps*), XP_020643852.1.

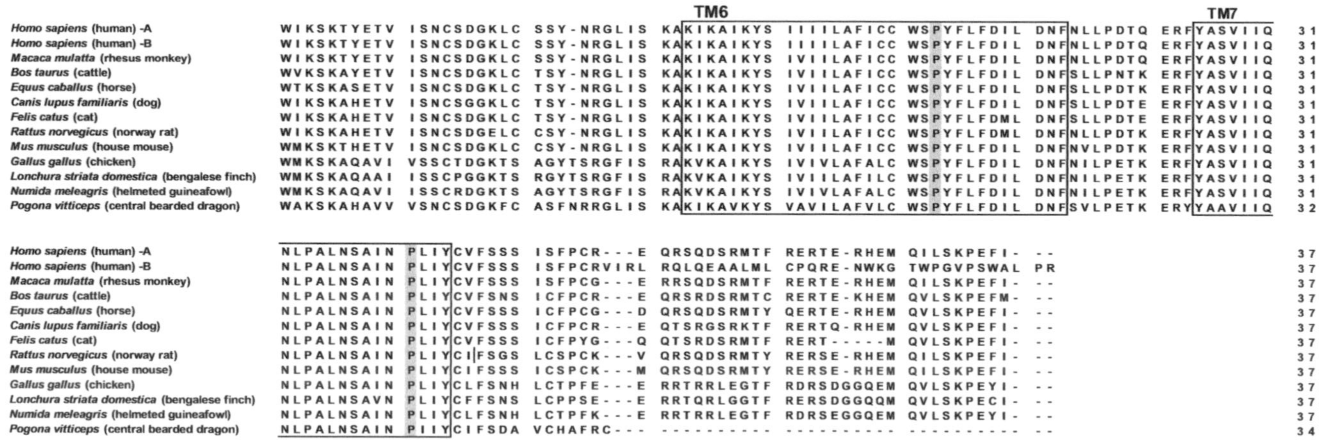

B

	Human-A	Human-B	Monkey	Cattle	Horse	Dog	Cat	Rat	Mouse	Chicken	Finch	Bearded dragon
Human-A	100	93	97	91	92	88	91	89	89	74	75	76
Human-B		100	90	86	85	83	86	83	83	71	72	76
Monkey			100	92	92	88	91	91	91	75	76	77
Cattle				100	93	91	93	89	89	77	77	79
Horse					100	90	93	89	89	75	76	78
Dog						100	93	87	86	75	75	78
Cat							100	89	87	76	77	77
Rat								100	97	75	76	78
Mouse									100	76	76	78
Chicken										100	90	80
Finch											100	77
Bearded dragon												100

Fig. 2—Cont'd

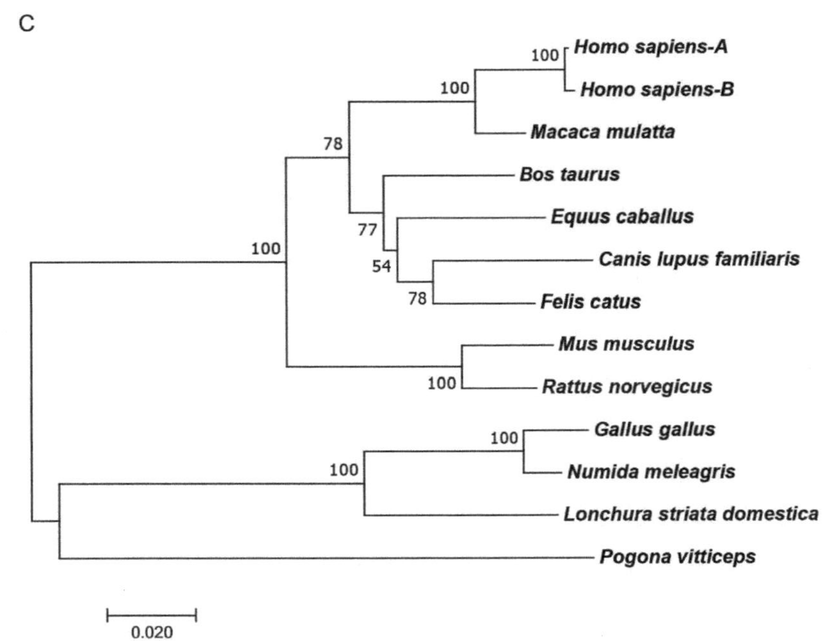
C
Homo sapiens-A
Homo sapiens-B
Macaca mulatta
Bos taurus
Equus caballus
Canis lupus familiaris
Felis catus
Mus musculus
Rattus norvegicus
Gallus gallus
Numida meleagris
Lonchura striata domestica
Pogona vitticeps
100
100
78
77
54
78
100
100
100
100
0.020

Fig. 2—Cont'd

Immunohistochemical assessments showed that both isoforms of NPSR1 are expressed in bronchus, colon, uterus, esophagus, spleen, kidney, pancreas, prostate, and breast. Isoform B, but not isoform A, is also expressed in stomach and small intestine. Compared with the faint staining in control samples, strong positive immunostaining of NPSR1-B can be found in smooth muscle and epithelial cells in asthmatic airways from bronchial biopsies, suggesting that isoform B might be more important in the pathogenesis of asthma.[16,29]

NPSR1 is also expressed in human leukocytes.[33] Both *NPSR1-A* and *NPSR1-B* transcripts are detected in monocytes and eosinophils. The mRNA expression level of isoform A in peripheral blood mononuclear cells is similar as in bronchial epithelial and smooth muscle cells, while the expression of isoform B in peripheral blood mononuclear cells is three times higher than in bronchial epithelial cells but lower than in bronchial smooth muscle cells. In addition, NPSR1-B is predominantly expressed in $CD4^+$ T cells. It is interesting to note that the expression of *NPSR1* and *NPS* in peripheral blood mononuclear cells can be upregulated by lipopolysaccharide stimulation.[33] NPSR1 expression is also observed in macrophages and eosinophils of human sputum samples by qualitative immunocytochemical analysis.[33]

In the mouse, expression profiling showed that *Npsr1* mRNA is expressed at high levels in hypothalamus, cerebral cortex, hippocampus, striatum, whole eye (including retina), lower levels in thymus and brainstem, and low levels in olfactory bulb and cerebellum.[34] Little or no expression is detected in the following tissues studied: pituitary gland, pancreas (including isolated islets of Langerhans), adrenal gland, thyroid/parathyroid gland, aorta, heart, lung, trachea, brown and white adipose tissue, liver, skeletal muscle, kidney, gall bladder, large and small intestine, stomach, urinary bladder, ovary, testes, uterus, skin, and spleen.[34]

In the rat brain, *Npsr1* is shown to be widely expressed in various regions. For example, *Npsr1* mRNA is expressed in areas involved in olfactory processing such as anterior olfactory nucleus.[5,12] In the input and output regions of the hippocampal formation including the parahippocampal regions, the lateral entorhinal cortex, and the retrosplenial agranular cortex, strong expression of *Npsr1* mRNA is also shown.[12] Finally, *Npsr1* mRNA is also expressed in the thalamus and hypothalamus, regions that are related to anxiety, sleep, and energy homeostasis.[5,12] The distribution of *Npsr1* transcripts in mouse brain is partially conserved between mouse and rat.[35] These results indicate NPSR1 might play very important roles in anxiety, arousal, energy balance, learning, and memory. Physiological experiments summarized below confirmed these suggestions.

3. Physiology of NPSR1

3.1 NPSR1 and immunoregulation

When NPSR1 target genes were analyzed, genes with ontology terms related to immune response and chemotaxis are overrepresented,[36] suggesting potential functional roles of NPSR1 in immunoregulation. Immunohistochemical analysis using inflammatory bronchial samples from classical ovalbumin-induced mouse model showed that NPSR1-A is highly expressed in the epithelial cells of the bronchioles and faintly expressed in alveolar macrophages in the lung tissue samples of naive and unsensitized but ovalbumin-challenged mice.[33] However, in the ovalbumin-sensitized mice, its expression is decreased in epithelial cells of the bronchioles and increased in alveolar macrophages compared with the control sample. These data suggest that NPSR1 might modulate macrophage immune responses.[33]

Peripheral blood mononuclear cells express both *NPS* and *NPSR1* mRNA.[33] Stimulation of peripheral blood mononuclear cells by lipopolysaccharide upregulates the expression of both *NPS* and *NPSR1* mRNA at 16 h posttreatment, while T-cell activation by anti-CD3 and anti-CD28 antibodies decreases the expression of *NPS* and *NPSR1* mRNA at 28 h after treatment. These results indicate that NPS may activate NPSR by an autocrine or paracrine mechanism.[33] In addition to lipopolysaccharide, interferon-γ and tumor necrosis factor-α also increase *NPSR1* mRNA levels in human mononuclear cells.[37,38]

Eosinophils from patients with severe asthma or high serum IgE levels have increased NPSR1 expression.[39] NPS also induces different signaling in these eosinophils, increasing Ca^{2+} but not cAMP in patients with high serum IgE but increasing cAMP and not Ca^{2+} in normal subjects.[39] NPSR activation increases eosinophil adhesion and migration, potentially contributing to the pathogenesis of asthma.[39]

3.2 NPSR1 and anxiety

Npsr1 mRNA is expressed in stress-related regions in brain such as the amygdala and hypothalamus. Therefore, NPSR1 might participate in regulating anxiety response. Xu et al. did a series of anxiety animal experiments at different conditions including open field, elevated plus maze, light–dark box, and marble-burying experiments.[5] In open field, the number of mice entering the central area in the first 10 min is increased after injection with

NPS. In the light–dark box, more mice spend longer time in the light area after 0.03–3 nmol NPS injection. With the elevated plus maze test, mice injected with 0.1 and 1 nmol NPS stay significantly longer time on the open arms and exhibit a higher number of transitions from closed to the open arms. In marble-burying experiment, mice injected with saline bury about half of glass marbles during the 30-min observation period, while NPS-injected mice bury significantly less marbles in a dose-dependent manner. Similar results were reported by Rizzi et al. in elevated plus maze experiment.[8]

Defensive burying test is an assay that can measure fear-like responses in stressful environmental conditions. Vitale et al. designed a rectangular Plexiglas burying apparatus with a bottom covered with fine sawdust and a shock probe set inside.[9] Then control and NPS-injected rats were placed into the apparatus separately. The results showed that control rats spend about 103 s to cover the probe with bedding material after shock and the height of buried bedding averages about 5.5 cm. But rats injected with 1 nmol NPS spend 47% less time to cover the probe, and the height of buried bedding is 45% lower than control rats. These studies indicate that NPSR1 plays an important role in reducing anxiety.[9]

In addition, Jüngling et al. found NPS can activate NPSR1 in presynaptic neurons to increase glutamatergic transmission to intercalated GABAergic neurons in the amygdala to potentially regulate conditioned fear. NPS injection into mouse amygdala can decrease conditioned fear response, whereas injection of NPS receptor antagonist exerts the opposite effect.[40]

In an elegant study examining the NPS system in rodents bred for high anxiety-related behavior vs low anxiety-related behavior, it was shown that the high anxiety-related behavior phenotype is associated with lower *Npsr1* expression.[41] Functional analysis of a synonymous SNP identified in the coding region of high anxiety-related behavior rodents showed it leads to higher cAMP response to NPS stimulation,[41] similar to the functional properties of human NPSR1 N107I (see Section 4).

3.3 NPSR1 and arousal and locomotor activity

As *Npsr1* mRNA is expressed in thalamic midline nuclei, it may also be involved in promoting arousal, reducing sleep, and increasing locomotor activity.[5,6] Xu et al. treated mice intracerebroventricularly with 0.1 and 1 nmol NPS and found significantly increased mouse locomotor activity.

A set of electrodes were implanted to record electroencephalograms and electromyograms. Saline-treated mice spend 45% of the first hour in wakefulness, whereas mice injected with 0.1 and 1 nmol NPS spend 69% and 87% of the first hour in wakefulness, respectively.[5] Interestingly, the increase in wakefulness during the first hour after NPS injection is followed by an increased amount of non-REM (rapid eye movement) sleep at the second and the fourth hour of experiment. Similar results were also obtained in several other studies.[8,11,42,43]

3.4 NPSR1 regulation of learning and memory

In the input and output regions of the hippocampal formation, as well as parahippocampal regions such as subiculum, lateral entorhinal cortex, and retrosplenial cortex, areas of the brain critical for regulating learning and memory, the expression of *Npsr1* mRNA is abundant.[5,12] Han et al. conducted a Morris water maze experiment in order to test the spatial learning and memory in mice.[14] Mice were trained for 4 days to find a submerged platform in a circular pool filled with milky water. NPS-injected mice have shorter latencies in finding the platform throughout the 4-day training period than those treated with artificial cerebrospinal fluid. And the day after training days, NPS-injected mice spend significantly longer time in the target quadrants than the control mice.[14] Knockout experiments showed that the NPS system is required for memory formation, since *Npsr1* knockout mice have deficits in memory functions.[44] NPS enhances long-term memory by working during the consolidation phase.[44]

Selective *N*-methyl-D-aspartate receptor antagonist MK801 can induce spatial memory impairment, and administration of NPS mitigates the impairment induced by MK801.[14] In male Sprague–Dawley rats, NPS has also been shown to mitigate impaired spatial memory caused by REM sleep deprivation and restore phosphorylation of cAMP-response element-binding protein.[45] These result suggests that NPS can improve spatial memory and mitigate spatial memory impairment in rodents.

3.5 NPSR1 regulation of food intake and energy balance

NPS has been shown to be a potent inhibitor of food intake. In Long–Evans rats, central injection of NPS inhibits voluntary and fasting-induced food intake to a similar degree as equimolar doses of corticotropin-releasing hormone.[46] NPS does not affect concentrations of leptin, ghrelin, and insulin, important central regulators of food intake, suggesting that the anorexic effect of NPS is independent of leptin, ghrelin, and insulin.[46] In Wistar rats, injection of NPS into the paraventricular nucleus inhibits food intake dose dependently.[11] Similar anorexigenic effect of NPS was also observed in chicks, with increased c-Fos reactivity in paraventricular nucleus,[13] where receptors critical for energy homeostasis, such as the melanocortin-4 receptor, are highly expressed.[47] Indeed, in Wistar male rats, injection of NPS directly into the paraventricular nucleus inhibits palatable food intake potently, whereas injection into the lateral hypothalamus produces a smaller and shorter-lasting decrease in food intake.[48] The effects of NPS are blocked by NPSR1 antagonist [D-Cys(*t*Bu)^5]NPS but not by a corticotropin-releasing hormone receptor antagonist, suggesting that the anorexic effect of NPS is mediated by NPSR1.[48]

However, there is also a report that NPS increases food intake. Niimi reported that intracerebroventricular injection of 1 nmol NPS can significantly increase food intake within 2 h in male Sprague–Dawley rats, with c-Fos activation in multiple nuclei, including paraventricular nucleus, dorsomedial nucleus, and lateral hypothalamic area, especially in orexinergic neurons.[10] In chicks, c-Fos activation was observed in paraventricular nucleus but not in ventromedial hypothalamus and lateral hypothalamic area.[13] The reason for the discrepancy between these studies is not clear.

4. Pharmacology of NPSR1

4.1 NPS structure–function

Using cells expressing recombinant hNPSR1, Roth et al. studied the effects of N- and C-terminal regions of hNPS on NPSR1 activation by progressive deletion of residues from both sides.[49] They found that the potency and efficacy of hNPS-(1–10) are the same as the full-length NPS. The activity will decrease with further truncations of the fragment until complete loss of activity in hNPS-(1–5) and hNPS-(1–4). Deletion of the N-terminal serine, hNPS-(2–20), results in a significant reduction in the potency of peptide; further deletion, from hNPS-(3–20) to hNPS-(10–20), results in total loss of activity.[49]

Bernier et al. did a similar study and showed the C-terminal truncated fragment hNPS-(1–13) maintains the same efficacy as full-length NPS but activity of hNPS-(1–10) is 20 times lower than full-length NPS,[50] different from data obtained by Roth and colleagues.[49] Leonard et al. showed that the C-terminal glutamine-truncated mouse NPS, NPS-(1–19), with high affinity for mouse NPSR1 and potently activates intracellular Ca^{2+} accumulation in vitro, is potent in vivo in three tests (four-plate test, elevated zero maze, and stress-induced hyperthermia).[7]

Alanine scanning of hNPS, systematic alanine replacement of each residue of the peptide, was used to delineate the contribution of amino acid side chain to the biological effect. Thus, 19 Ala-substituted hNPS analogs were synthesized and tested for their ability to activate NPSR1. [Ala^3], [Ala^4], and [Ala^7] hNPS produced weak or no stimulatory effect on Ca^{2+} mobilization in HEK293 cells expressing hNPSR1. Substitution of Arg3 with Ala decreases the maximal response by half. Ala substitutions from other positions show no loss of agonist effect.[49] In another similar experiment, substitution of residues 2, 3, and 4 with Ala results in substantial to complete inactivation of the peptide, as does the G7A mutation.[50]

The positions 2, 3, 4, and 5 on N-terminus of NPS play important roles in activation of the receptor.[51–54] NPS position 2 (Phe) is of vital importance in both receptor binding and activation. Lipophilicity of position 2 is more important than its aromaticity, and both size and distance of the chemical moiety are crucial for its biological activity.[51] The receptor pocket allocating the side chain of position 3 (Phe) can accommodate slightly larger side chains. An aliphatic amino acid with a linear three carbon atom long side chain is enough to activate NPSR. Almost all the amino acid replacements of position 4 (Asn) produce weak or no activation. This result indicates Asn4 side chain is critical for NPS bioactivity.[52] As for position 5 (Gly), [D-Cys(*t*Bu)^5]NPS, [D-Val^5]NPS, [D-Trp^5]NPS, [D-Met^5]NPS, and [D-Phe^5]NPS behave as pure and selective NPSR antagonists, and [D-Leu^5]NPS and [D-Cys^5]NPS behave as partial agonists.[53,54] In 2010, two compounds, NPSR-QA1 and NPSR-PI1, were identified as antagonists of NPSR1,[55,56] both of which were further tested and verified by Camarda et al.[57]

Recently, using peptide welding technology, a tetrabranched NPS analog was synthesized and shown to have higher potency in vitro and longer-lasting action in vivo than native NPS.[58]

Several studies reported identification of small molecule NPSR1 antagonists, although no agonist has been identified (see ref. 59 for a recent comprehensive review). Two nonpeptidic bicyclic piperazines, SHA 66 and SHA 68, were identified as NPSR1 antagonists.[60] These compounds can inhibit NPS-induced Ca^{2+} mobilization both in vivo and in vitro. Importantly, SHA 68 can reach pharmacologically relevant levels in plasma and brain after intraperitoneal injection, suggesting that SHA 68 can cross the blood–brain barrier. The antagonistic activity of SHA 68 appears to be specific as it does not affect signaling of other unrelated GPCRs. Because peripheral administration can antagonize NPS-induced activities, SHA 68 should be a useful tool to study the physiological characteristics and pharmacological parameters of NPS–NPSR1 system in the central nervous system.[60] SHA 68 has also been shown to have pharmacological chaperone activity, rescuing intracellularly retained mutant receptors to the plasma membrane.[61]

Recently, an imidazopyridine, NCGC00185684, was identified as a selective and brain penetrant antagonist of NPSR1 with high affinity.[62] Regions such as central amygdala or nucleus accumbens shell and core in brain are involved in regulation of alcohol intake, and alcohol intake can lead to ERK phosphorylation. NCGC00185684 can preferentially block alcohol-induced ERK phosphorylation over intracellular calcium and cAMP responses to NPS. This finding provided initial evidence that NPSR1 may become a therapeutic target in alcoholism.[62]

4.2 NPSR1 signaling pathways

NPSR1 is coupled to both Gq and Gs proteins.[5,17,63] In both HEK293T and CHO cell lines, NPS can dose-dependently increase intracellular Ca^{2+} and cAMP concentrations with EC_{50} of about 5–10 nM.[5] In Gq pathway, NPS binding to NPSR1 leads to activation of Gq–phospholipase C and cleavage of phospholipid phosphatidylinositol 4,5-bisphosphate to inositol 1,4,5-trisphosphate (IP3) and diacylglycerol. Then IP3 activates IP3 receptor on the endoplasmic reticulum membrane and releases Ca^{2+} from the endoplasmic reticulum. In primary hippocampal neurons expressing NPSR1, store-operated calcium channels are also activated, allowing entry of extracellular calcium into the cells.[64]

In the Gs pathway, activation of NPSR1 results in activating adenylate cyclase, increased synthesis of cAMP, and activation of protein kinase A. In human, NPSR1-B signals through the same pathways and regulates the same genes as NPSR1-A, but it is interesting that NPSR1-A induces stronger signaling effects than NPSR1-B,[65] although an earlier study showed that NPSR1-A and NPSR1-B have similar pharmacological profiles.[63] No constitutive activity is observed with NPSR1.[17]

In addition, like other GPCRs, NPSR1 activation also leads to activation of extracellular signal-regulated kinase 1/2 (ERK1/2).[62,63,65,66] In HEK293 cells stably expressing NPSR1, dose-dependent stimulation of ERK1/2 phosphorylation was observed, and in Colo205 human colon cancer cells, NPS stimulates cell proliferation.[63]

As mentioned earlier, both NPS and NPSR1 associate with asthma and the expression of NPSR1 in inflammatory tissue is increased. Matrix metallopeptidase 10 (MMP10) and tissue inhibitor of metallopeptidase inhibitor 3 (TIMP3) are also asthma-related genes. The strong expression of both MMP10 and TIMP3 is found to colocalize with NPSR1-A in bronchial epithelium. In NPSR1-A overexpressing cells stimulated with NPS the expression of MMP10 is also increased.[36] This result suggests NPS–NPSR1 signaling pathway leads to upregulation of TIMP3 and MMP10 and may be associated with the pathogenesis of asthma.

Recently, biased signaling has also been reported for NPSR1. Full-length NPS has similar potencies at the Gq and Gs pathways, therefore acting as a balanced agonist. By measuring signaling at both Gq and Gs pathways, it was shown that human NPS(1–10), lacking the C-terminal 10 amino acids, is a biased agonist activating Gq pathway more potently.[66] When ERK1/2 activation was studied, full-length NPS activates ERK1/2 through both Gq- and Gs-mediated signaling, whereas NPS(1–10) activates ERK1/2 through Gq-mediated signaling only.[66] A high-affinity brain penetrant NPSR1 antagonist, NCGC00185684, preferentially blocks ERK1/2 activation over intracellular cAMP or calcium accumulation.[62] A modified tetrapeptide NPS analog (Compound 4), biased toward the Gq–calcium mobilization pathway in vitro, has diminished locomotor stimulation yet retains activity in memory formation and anxiolytics in vivo.[67] Biased ligands like this would represent a better therapeutics with less side effects.

5. Pathophysiology of NPSR1

Data from Exome Aggregation Consortium database showed that NPSR1 is highly polymorphic in the coding region, with a large number of naturally occurring mutations or polymorphisms (Fig. 1).[68,69] The functional implications of these mutations and the associated clinical phenotypes are not known at present. We summarize below the SNPs that are known to be associated with human illness.

5.1 *NPSR1* polymorphisms and asthma

Asthma is associated with changes in the expression of the two isoforms of *NPSR1* (see Section 2.2). Importantly, several SNPs and haplotypes of

NPSR1 have been identified to be associated with asthma in several populations.[16,19–25] Most of the SNPs associated with asthma are intronic variants.

Following the initial pioneering report of association of *NPSR1* SNPs with asthma in three cohorts from Finland and Canada,[16] two additional studies quickly confirmed the association in Caucasian populations. Seven SNPs of *NPSR1* gene, rs323917 (C/G), rs323922 (G/C), rs324377 (C/A), SNP546333 (G/A), rs324384 (T/C), rs324396 (C/T), and rs740347 (G/C), were genotyped in a German population and Western European populations.[24,25] In the German population, rs324384 was excluded from analysis because it significantly deviates from Hardy–Weinberg equilibrium. SNP546333 and haplotype (CCAACC) are associated with increased risk for asthma and concomitant asthma and bronchial hyperresponsiveness, and rs740347 is associated with asthma, asthma in combination with bronchial hyperresponsiveness and elevated serum IgE levels.[24] For the Western European population, haplotype 1 (CGCGCTG), haplotype 5 (CCAGTCC), and haplotype 6 (CGCGTCG) are significantly associated with sensitization. Moreover, haplotype 1 (CGCGCTG) and haplotype 5 (CCAGTCC) are significantly associated with childhood allergic asthma.[25] Similar study was conducted in an Italian population by family-based association analyses.[21] Only SNP546333 and rs740347 are associated with elevated IgE levels. Haplotype (CCAATCC) is less frequent in individuals with elevated IgE.[21]

Although a study in Korean population failed to confirm the association of NPSR1 with asthma,[70] in a Chinese population, rs324981 (T/A) and rs324987 (C/T) are associated with a significantly lower risk of airway hyperresponsiveness to methacholine.[23] A/A homozygote of rs324981 was shown as a protective factor for airway hyperresponsiveness to methacholine in Chinese population.[23]

In 2007, a comprehensive investigation reported the polymorphisms of *NPSR1* gene in two populations, European-American children from the Childhood Asthma Management Program and Hispanic children from the Central Valley of Costa Rica.[22] Of the genotyped SNPs in five genes, only three SNPs in *NPSR1* are associated with asthma in both groups (rs2609234, rs714588, rs1379928). Interestingly, the direction of effects of these associated alleles are opposite in the two groups,[22] potentially due to different linkage disequilibrium patterns or failure to incorporate potential gene–gene and/or gene–environment interactions.[71]

Vergara et al. genotyped six SNPs of *NPSR1*, rs323917 (C/G), rs323922 (C/G), rs324377 (A/C), Hopo546333 (G/A), rs740347 (C/G), and rs324981 (A/T) in a population of Colombia.[19] They showed that allele

A of SNP546333 is protective for asthma and allele C of rs740347 is negatively associated with both asthma and high IgE levels. The haplotype (CGCGGT) is positively associated with asthma as well as high IgE levels, whereas haplotype (CCAGGT) only shows significant association with high IgE levels but not with asthma.[19]

Castro-Giner et al. genotyped *NPSR1* SNPs of European Community Respiratory Health Survey participants from 13 countries.[20] They identified six and eight SNPs nominally significantly associated with asthma and atopy, respectively. Only three SNPs (rs184448, rs324957, rs324981) associated with asthma, and just one SNP (rs324396) with atopy was confirmed after correction for country, body mass index, smoking status, and sex.[20]

5.2 *NPSR1* polymorphisms and anxiety and psychiatric disorders

NPS is known to cause arousal and have anxiolytic effects in rodents. In human *NPSR1* gene, SNP rs324981 A>T results in NPSR1 N107I variant.[63] In a Japanese population, WT *NPSR1* carriers are found underrepresented in panic disorder patients compared to controls, but the T allele increases the risk of developing panic disorder only in males.[72]

Caffeine is known to provoke panic disorder.[73] Caffeine can downregulate NPS and NPSR expression levels, decreasing *Nps* mRNA levels in brainstem, and increasing *Npsr1* mRNA levels in the hypothalamus and brainstem, suggesting that some of the behavioral effects of caffeine (such as panic-provoking) might be mediated by the NPS–NPSR system,[74] in addition to the well-known action of caffeine as an antagonist of adenosine receptors. Caffeine attenuates the stimulatory effects of NPS on locomotion and rearing in mice, which is blocked by specific adenosine A_{2A} receptor antagonist (ZM241385) but not by specific adenosine A_1 receptor antagonist (CPT), suggesting the involvement of adenosine A_{2A} receptor in modulating the stimulatory effect of NPS in locomotion.[75] In an emotionally relevant environmental stimuli test, 24 emotionally threatening unpleasant, 24 pleasant, and 24 neutral images were used to test 124 healthy subjects.[76] When challenged with caffeine, NPSR1 N107I carriers show a significantly increased startle magnitude in response to neutral stimuli and significantly decreased startle response to unpleasant stimuli compared with WT and heterozygotes.[76]

By assessing affective and anxiety disorders of an Estonian sample of the European Youth Heart Study, Laas et al. found females with WT NPSR1 show five times higher odds of having anxiety disorder than females with

NPSR1 N107I and heterozygote in negative family environment. Females with WT NPSR1 attempt to suicide more frequently, which is further increased by adverse family environment.[77]

NPSR1 gene is also associated with obsessive–compulsive disorder. Obsessive–compulsive disorder patients with WT *NPSR1* genotype have significantly earlier onset than N107I and heterozygote carriers.[78] The interaction between SNP rs6265 of brain-derived neurotrophic factor and NPSR1 N107I increases the risk of depression in asthmatic patients.[79]

5.3 *NPSR1* polymorphisms and other diseases

SNPs of *NPSR1* gene are also related to other diseases and physiological response. Haplotype 3 (CGCGCCG) is associated with allergic rhinoconjunctivitis in Western European countries.[25] Gene–environment interaction has also been observed. SNPs of *NPSR1* gene are found to interact with regular farm animal contact in development of allergic symptoms in European children.[37]

NPSR1 SNPs are also associated with rheumatoid arthritis susceptibility. rs324987 and rs10263447 are significantly related to rheumatoid arthritis in Epidemiological Investigation of Rheumatoid Arthritis case–control study and rs740347 in a Spanish population.[26,28] Robledo et al. found *NPSR1* haplotypes CGCCGGC and CGCTGTG are associated with rheumatoid arthritis susceptibility and haplotypes CCACCGC is associated with rheumatoid arthritis protection.[28]

NPSR1 SNPs are associated with susceptibility to gastrointestinal disorders. D'Amato et al. found haplotypes of *NPSR1* gene show strong association with IBD in Italian, Swedish, and Finnish populations.[27] *NPSR1* haplotype H2 is associated with predisposing effect in Crohn's disease, and haplotype H8 is associated with protective effect in ulcerative colitis according to 1664 IBD patients and 826 controls from Italy, Sweden, and Finland.[27] *NPSR1* functional haplotype rs2530547C–rs324981A–rs727162C is significantly associated with reduced risk of IBD.[80]

5.4 Functional studies of naturally occurring NPSR1 variants

Several studies investigated the most common variant, N107I. Functional studies showed that N107I has higher expression than the WT NPSR1.[50,63] (In Anedda et al.'s paper,[80] the variant N107I together with S241R was used as the WT receptor; hence data in this paper are very difficult to interpret.) Because the total expression is not increased,[50] the increased cell surface expression represents more efficient trafficking of N107I compared with the WT receptor. This represents a new type of gain-of-function mutation, different from previously identified types of gain-of-function mutations such as constitutive activation, increased sensitivity, decreased specificity to ligands, and decreased desensitization.[81] This also suggests that the WT NPSR1 is not efficiently transported to the cell surface, similar to several other GPCRs.[31]

Compared with WT NPSR1 (N107), N107I variant has 10-fold higher potency in response to NPS stimulation when intracellular cAMP and Ca^{2+} concentrations were measured.[63] When cell lines expressing similar levels of WT NPSR1-A and NPSR1-A N107I were used for dose–response experiments to measure intracellular Ca^{2+} concentrations, EC_{50}s of 12.4 and 1.4 nM were obtained for WT and N107I NPSR1-A, respectively. EC_{50}s of 31.9 and 3.5 nM were obtained for WT and N107I NPSR1-A, respectively, when intracellular cAMP accumulation was measured.[63]

6. Summary

GPCRs are the largest family of membrane proteins in human. Because they regulate almost all physiological processes, GPCRs have always been important targets for drug development. Asthma is one of the most common chronic diseases, and NPSR1, a member of family A GPCRs, shows significant association with asthma. It is very important to study the NPSR1 mutants and identify ligands for developing new drugs for asthma. Although there are some studies on the functions and SNPs of NPSR1, our knowledge about the relationship between NPSR1 and asthma is still limited. More studies on NPSR1 are needed to explore the therapeutic potential of NPSR1 for asthma in the future.

Acknowledgment

Z.-R.Z. was supported by a fellowship from China Scholarship Council of the People's Republic of China.

References

1. Vos T, Allen C, Arora M, et al. Global, regional, and national incidence, prevalence, and years lived with disability for 310 diseases and injuries, 1990–2015: a systematic analysis for the global burden of disease study 2015. *Lancet*. 2016;388:1545–1602.
2. Martinez FD. Genes, environments, development and asthma: a reappraisal. *Eur Respir J*. 2007;29:179–184.
3. Sato S, Shintani Y, Miyajima N, Yoshimula K. *Novel G Protein-Coupled Receptor Protein and DNA Thereof*. 2002. European Patent Office, WO 02/31145 A1.

4. Reinscheid RK. Phylogenetic appearance of neuropeptide S precursor proteins in tetrapods. *Peptides*. 2007;28:830–837.
5. Xu YL, Reinscheid RK, Huitron-Resendiz S, et al. Neuropeptide S: a neuropeptide promoting arousal and anxiolytic-like effects. *Neuron*. 2004;43:487–497.
6. Reinscheid RK, Xu YL. Neuropeptide S and its receptor: a newly deorphanized G protein-coupled receptor system. *Neuroscientist*. 2005;11:532–538.
7. Leonard SK, Dwyer JM, Rizzo SJS, et al. Pharmacology of neuropeptide S in mice: therapeutic relevance to anxiety disorders. *Psychopharmacology*. 2008;197:601–611.
8. Rizzi A, Vergura R, Marzola G, et al. Neuropeptide S is a stimulatory anxiolytic agent: a behavioural study in mice. *Br J Pharmacol*. 2008;154:471–479.
9. Vitale G, Filaferro M, Ruggieri V, et al. Anxiolytic-like effect of neuropeptide S in the rat defensive burying. *Peptides*. 2008;29:2286–2291.
10. Niimi M. Centrally administered neuropeptide S activates orexin-containing neurons in the hypothalamus and stimulates feeding in rats. *Endocrine*. 2006;30:75–79.
11. Smith KL, Patterson M, Dhillo WS, et al. Neuropeptide S stimulates the hypothalamo-pituitary-adrenal axis and inhibits food intake. *Endocrinology*. 2006;147:3510–3518.
12. Xu YL, Gall CM, Jackson VR, Civelli O, Reinscheid RK. Distribution of neuropeptide S receptor mRNA and neurochemical characteristics of neuropeptide S-expressing neurons in the rat brain. *J Comp Neurol*. 2007;500:84–102.
13. Cline MA, Godlove DC, Nandar W, Bowden CN, Prall BC. Anorexigenic effects of central neuropeptide S involve the hypothalamus in chicks (*Gallus gallus*). *Comp Biochem Physiol A Mol Integr Physiol*. 2007;148:657–663.
14. Han RW, Yin XQ, Chang M, Peng YL, Li W, Wang R. Neuropeptide S facilitates spatial memory and mitigates spatial memory impairment induced by N-methyl-D-aspartate receptor antagonist in mice. *Neurosci Lett*. 2009;455:74–77.
15. Acevedo N, Ezer S, Merid SK, et al. Neuropeptide S (*NPS*) variants modify the signaling and risk effects of NPS receptor 1 (*NPSR1*) variants in asthma. *PLoS One*. 2017;12: e0176568.
16. Laitinen T, Polvi A, Rydman P, et al. Characterization of a common susceptibility locus for asthma-related traits. *Science*. 2004;304:300–304.
17. Gupte J, Cutler G, Chen JL, Tian H. Elucidation of signaling properties of vasopressin receptor-related receptor 1 by using the chimeric receptor approach. *Proc Natl Acad Sci USA*. 2004;101:1508–1513.
18. Daley D, Lemire M, Akhabir L, et al. Analyses of associations with asthma in four asthma population samples from Canada and Australia. *Hum Genet*. 2009;125: 445–459.
19. Vergara C, Jimenez S, Acevedo N, et al. Association of G-protein-coupled receptor 154 with asthma and total IgE in a population of the Caribbean coast of Colombia. *Clin Exp Allergy*. 2009;39:1558–1568.
20. Castro-Giner F, de Cid R, Gonzalez JR, et al. Positionally cloned genes and age specific effects in asthma and atopy: an international population-based cohort study (ECRHS). *Thorax*. 2010;65:124–131.
21. Malerba G, Lindgren C, Xumerle L, et al. Chromosome 7p linkage and *GPR154* gene association in Italian families with allergic asthma. *Clin Exp Allergy*. 2007;37: 83–89.
22. Hersh CP, Raby BA, Soto-Quirós ME, et al. Comprehensive testing of positionally cloned asthma genes in two populations. *Am J Respir Crit Care Med*. 2007;176: 849–857.
23. Feng Y, Hong XM, Wang L, et al. G protein-coupled receptor 154 gene polymorphism is associated with airway hyperresponsiveness to methacholine in a Chinese population. *J Allergy Clin Immunol*. 2006;117:612–617.
24. Kormann MSD, Carr D, Klopp N, et al. G-protein-coupled receptor polymorphisms are associated with asthma in a large German population. *Am J Respir Crit Care Med*. 2005;171:1358–1362.

25. Melén E, Bruce S, Doekes G, et al. Haplotypes of G protein-coupled receptor 154 are associated with childhood allergy and asthma. *Am J Respir Crit Care Med.* 2005;171:1089–1095.
26. D'Amato M, Zucchelli M, Seddighzadeh M, et al. Analysis of neuropeptide S receptor gene (*NPSR1*) polymorphism in rheumatoid arthritis. *PLoS One.* 2010;5:e9315.
27. D'Amato M, Bruce S, Bresso F, et al. Neuropeptide S receptor 1 gene polymorphism is associated with susceptibility to inflammatory bowel disease. *Gastroenterology.* 2007;133:808–817.
28. Robledo G, González-Gay MA, Fernández-Gutiérrez B, et al. *NPSR1* gene is associated with reduced risk of rheumatoid arthritis. *J Rheumatol.* 2012;39:1166–1170.
29. Vendelin J, Pulkkinen V, Rehn M, et al. Characterization of *GPRA*, a novel G protein-coupled receptor related to asthma. *Am J Respir Cell Mol Biol.* 2005;33:262–270.
30. Tao YX. Inactivating mutations of G protein-coupled receptors and diseases: structure-function insights and therapeutic implications. *Pharmacol Ther.* 2006;111:949–973.
31. Tao YX, Conn PM. Chaperoning G protein-coupled receptors: from cell biology to therapeutics. *Endocr Rev.* 2014;35:602–647.
32. Pitti T, Manoj N. Molecular evolution of the neuropeptide S receptor. *PLoS One.* 2012;7:e34046.
33. Pulkkinen V, Majuri ML, Wang GY, et al. Neuropeptide S and G protein-coupled receptor 154 modulate macrophage immune responses. *Hum Mol Genet.* 2006;15:1667–1679.
34. Regard JB, Sato IT, Coughlin SR. Anatomical profiling of G protein-coupled receptor expression. *Cell.* 2008;135:561–571.
35. Clark SD, Duangdao DM, Schulz S, et al. Anatomical characterization of the neuropeptide S system in the mouse brain by *in situ* hybridization and immunohistochemistry. *J Comp Neurol.* 2011;519:1867–1893.
36. Vendelin J, Bruce S, Holopainen P, et al. Downstream target genes of the neuropeptide S–NPSR1 pathway. *Hum Mol Genet.* 2006;15:2923–2935.
37. Bruce S, Nyberg F, Melén E, et al. The protective effect of farm animal exposure on childhood allergy is modified by *NPSR1* polymorphisms. *J Med Genet.* 2009;46:159–167.
38. Sundman L, Saarialho-Kere U, Vendelin J, et al. Neuropeptide S receptor 1 expression in the intestine and skin—putative role in peptide hormone secretion. *Neurogastroenterol Motil.* 2010;22:79–87. e30.
39. Ilmarinen P, James A, Moilanen E, et al. Enhanced expression of neuropeptide S (NPS) receptor in eosinophils from severe asthmatics and subjects with total IgE above 100 IU/ml. *Peptides.* 2014;51:100–109.
40. Jüngling K, Seidenbecher T, Sosulina L, et al. Neuropeptide S-mediated control of fear expression and extinction: role of intercalated GABAergic neurons in the amygdala. *Neuron.* 2008;59:298–310.
41. Slattery DA, Naik RR, Grund T, et al. Selective breeding for high anxiety introduces a synonymous SNP that increases neuropeptide S receptor activity. *J Neurosci.* 2015;35:4599–4613.
42. Li W, Gao YH, Chang M, et al. Neuropeptide S inhibits the acquisition and the expression of conditioned place preference to morphine in mice. *Peptides.* 2009;30:234–240.
43. Ensho T, Nakahara K, Suzuki Y, Murakami N. Neuropeptide S increases motor activity and thermogenesis in the rat through sympathetic activation. *Neuropeptides.* 2017;65:21–27.
44. Okamura N, Garau C, Duangdao DM, et al. Neuropeptide S enhances memory during the consolidation phase and interacts with noradrenergic systems in the brain. *Neuropsychopharmacology.* 2011;36:744–752.

45. Zhao ZQ, Huang LQ, Wu HJ, et al. Neuropeptide S mitigates spatial memory impairment induced by rapid eye movement sleep deprivation in rats. *Neuroreport.* 2010;21:623–628.
46. Beck B, Fernette B, Stricker-Krongrad A. Peptide S is a novel potent inhibitor of voluntary and fast-induced food intake in rats. *Biochem Biophys Res Commun.* 2005;332:859–865.
47. Tao YX. The melanocortin-4 receptor: physiology, pharmacology, and pathophysiology. *Endocr Rev.* 2010;31:506–543.
48. Fedeli A, Braconi S, Economidou D, et al. The paraventricular nucleus of the hypothalamus is a neuroanatomical substrate for the inhibition of palatable food intake by neuropeptide S. *Eur J Neurosci.* 2009;30:1594–1602.
49. Roth AL, Marzola E, Rizzi A, et al. Structure activity studies on neuropeptide S: identification of the amino acid residues crucial for receptor activation. *J Biol Chem.* 2006;281:20809–20816.
50. Bernier V, Stocco R, Bogusky MJ, et al. Structure-function relationships in the neuropeptide S receptor. Molecular consequences of the asthma-associated mutation N107I. *J Biol Chem.* 2006;281:24704–24712.
51. Camarda V, Trapella C, Calò G, et al. Synthesis and biological activity of human neuropeptide S analogues modified in position 2. *J Med Chem.* 2008;51:655–658.
52. Camarda V, Trapella C, Guerrini R, et al. Structure–activity study at positions 3 and 4 of human neuropeptide S. *Bioorg Med Chem.* 2008;16:8841–8845.
53. Camarda V, Rizzi A, Ruzza C, et al. *In vitro* and *in vivo* pharmacological characterization of the neuropeptide S receptor antagonist [D-Cys (tBu) 5] neuropeptide S. *J Pharmacol Exp Ther.* 2009;328:549–555.
54. Guerrini R, Camarda V, Trapella C, et al. Synthesis and biological activity of human neuropeptide S analogues modified in position 5: identification of potent and pure neuropeptide S receptor antagonists. *J Med Chem.* 2008;52:524–529.
55. Melamed JY, Zartman AE, Kett NR, et al. Synthesis and evaluation of a new series of neuropeptide S receptor antagonists. *Bioorg Med Chem Lett.* 2010;20:4700–4703.
56. Trotter BW, Nanda KK, Manley PJ, et al. Tricyclic imidazole antagonists of the neuropeptide S receptor. *Bioorg Med Chem Lett.* 2010;20:4704–4708.
57. Camarda V, Ruzza C, Rizzi A, et al. *In vitro* and *in vivo* pharmacological characterization of the novel neuropeptide S receptor ligands QA1 and PI1. *Peptides.* 2013;48:27–35.
58. Ruzza C, Rizzi A, Malfacini D, et al. *In vitro* and *in vivo* pharmacological characterization of a neuropeptide S tetrabranched derivative. *Pharmacol Res Perspect.* 2015;3, e00108.
59. Ruzza C, Calò G, Di Maro S, et al. Neuropeptide S receptor ligands: a patent review (2005–2016). *Expert Opin Ther Pat.* 2017;27:347–362.
60. Okamura N, Habay SA, Zeng J, Chamberlin AR, Reinscheid RK. Synthesis and pharmacological *in vitro* and *in vivo* profile of 3-oxo-1, 1-diphenyl-tetrahydro-oxazolo [3, 4-a] pyrazine-7-carboxylic acid 4-fluoro-benzylamide (SHA 68), a selective antagonist of the neuropeptide S receptor. *J Pharmacol Exp Ther.* 2008;325:893–901.
61. Clark SD, Tran HT, Zeng J, Reinscheid RK. Importance of extracellular loop one of the neuropeptide S receptor for biogenesis and function. *Peptides.* 2010;31:130–138.
62. Thorsell A, Tapocik JD, Liu K, et al. A novel brain penetrant NPS receptor antagonist, NCGC00185684, blocks alcohol-induced ERK-phosphorylation in the central amygdala and decreases operant alcohol self-administration in rats. *J Neurosci.* 2013;33:10132–10142.
63. Reinscheid RK, Xu YL, Okamura N, et al. Pharmacological characterization of human and murine neuropeptide S receptor variants. *J Pharmacol Exp Ther.* 2005;315:1338–1345.

64. Erdmann F, Kügler S, Blaesse P, et al. Neuronal expression of the human neuropeptide S receptor NPSR1 identifies NPS-induced calcium signaling pathways. *PLoS One*. 2015;10:e0117319.
65. Pietras CO, Vendelin J, Anedda F, et al. The asthma candidate gene *NPSR1* mediates isoform specific downstream signalling. *BMC Pulm Med*. 2011;11:39.
66. Liao Y, Lu B, Ma Q, et al. Human neuropeptide S receptor is activated via a Gαq protein-biased signaling cascade by a human neuropeptide S analog lacking the C-terminal ten residues. *J Biol Chem*. 2016;291:7505–7516.
67. Clark SD, Kenakin TP, Gertz S, et al. Identification of the first biased NPS receptor agonist that retains anxiolytic and memory promoting effects with reduced levels of locomotor stimulation. *Neuropharmacology*. 2017;118:69–78.
68. Lek M, Karczewski KJ, Minikel EV, et al. Analysis of protein-coding genetic variation in 60,706 humans. *Nature*. 2016;536:285–297.
69. Karczewski KJ, Weisburd B, Thomas B, et al. The ExAC browser: displaying reference data information from over 60 000 exomes. *Nucleic Acids Res*. 2016;45:D840–D845.
70. Shin HD, Park KS, Park CS. Lack of association of *GPRA* (G protein-coupled receptor for asthma susceptibility) haplotypes with high serum IgE or asthma in a Korean population. *J Allergy Clin Immunol*. 2004;114:1226–1227.
71. Lin PI, Vance JM, Pericak-Vance MA, Martin ER. No gene is an island: the flip-flop phenomenon. *Am J Hum Genet*. 2007;80:531–538.
72. Okamura N, Hashimoto K, Iyo M, et al. Gender-specific association of a functional coding polymorphism in the neuropeptide S receptor gene with panic disorder but not with schizophrenia or attention-deficit/hyperactivity disorder. *Prog Neuropsychopharmacol Biol Psychiatry*. 2007;31:1444–1448.
73. Bourin M, Malinge M, Guitton B. Provocative agents in panic disorder. *Therapie*. 1995;50:301–306.
74. Lage R, Diéguez C, López M. Caffeine treatment regulates neuropeptide S system expression in the rat brain. *Neurosci Lett*. 2006;410:47–51.
75. Boeck CR, Martinello C, de Castro AA, et al. Blockade of adenosine A2A receptor counteracts neuropeptide-S-induced hyperlocomotion in mice. *Naunyn Schmiedebergs Arch Pharmacol*. 2010;381:153–160.
76. Domschke K, Klauke B, Winter B, et al. Modification of caffeine effects on the affect-modulated startle by neuropeptide S receptor gene variation. *Psychopharmacology*. 2012;222:533–541.
77. Laas K, Reif A, Akkermann K, et al. Interaction of the neuropeptide S receptor gene Asn107Ile variant and environment: contribution to affective and anxiety disorders, and suicidal behaviour. *Int J Neuropsychopharmacol*. 2014;17:541–552.
78. Lennertz L, Franke PE, Grabe HJ, et al. The functional coding variant Asn107Ile of the neuropeptide S receptor gene (*NPSR1*) influences age at onset of obsessive–compulsive disorder. *Int J Neuropsychopharmacol*. 2013;16:1951–1958.
79. Yang Y, Zhao M, Zhang Y, Shen X, Yuan Y. Correlation of *5-HTT*, *BDNF* and *NPSR1* gene polymorphisms with anxiety and depression in asthmatic patients. *Int J Mol Med*. 2016;38:65–74.
80. Anedda F, Zucchelli M, Schepis D, et al. Multiple polymorphisms affect expression and function of the neuropeptide S receptor (*NPSR1*). *PLoS One*. 2011;6:e29523.
81. Tao YX. Constitutive activation of G protein-coupled receptors and diseases: insights into mechanism of activation and therapeutics. *Pharmacol Ther*. 2008;120:129–148.

CHAPTER EIGHT

Prokineticins and their G protein-coupled receptors in health and disease

Yaguang Zhao[*,†,‡], Jiayu Wu[*,†,‡], Xinying Wang[*,†,‡], Hong Jia[*,†,‡], Dan-Na Chen[§,1], Jia-Da Li[*,†,‡,1]
[*]School of Life Sciences, Central South University, Changsha, China
[†]Hunan Key Laboratory of Animal Models for Human Diseases, Central South University, Changsha, China
[‡]Hunan Key Laboratory of Medical Genetics, Central South University, Changsha, China
[§]Department of Basic Medical Sciences, Changsha Medical University, Changsha, China
[1]Corresponding authors: e-mail address: chendanna007@gmail.com; lijiada@sklmg.edu.cn

Contents

Abstract

Prokineticins are two conserved small proteins (~8 kDa), prokineticin 1 (PROK1; also called EG-VEGF) and prokineticin 2 (PROK2; also called Bv8), with an N-terminal AVITGA sequence and 10 cysteines forming 5 disulfide bridges. PROK1 and PROK2 bind to two

Progress in Molecular Biology and Translational Science, Volume 161
ISSN 1877-1173
https://doi.org/10.1016/bs.pmbts.2018.09.006

highly related G protein-coupled receptors (GPCRs), prokineticin receptor 1 (PROKR1) and prokineticin receptor 2 (PROKR2). Prokineticins and their receptors are widely expressed. PROK1 is predominantly expressed in peripheral tissues, especially steroidogenic organs, whereas PROK2 is mainly expressed in the central nervous system and nonsteroidogenic cells of the testes. Prokineticins signaling has been implicated in several important physiological functions, including gastrointestinal smooth muscle contraction, circadian rhythm regulation, neurogenesis, angiogenesis, pain perception, mood regulation, and reproduction. Dysregulation of prokineticins signaling has been observed in a variety of diseases, such as cancer, ischemia, and neurodegeneration, in which prokineticins signaling seems to be a promising therapeutic target. Based on the phenotypes of knock-out mice, *PROKR2* and *PROK2* have recently been identified as causative genes for idiopathic hypogonadotropic hypogonadism, a developmental disorder characterized by impaired development of gonadotropin-releasing hormone neurons and infertility. In vitro functional studies with these disease-associated *PROKR2* mutations uncovered some novel features for this receptor, such as biased signaling, which may be used to understand GPCR signaling regulation in general.

Abbreviations

Aβ amyloid β
AD Alzheimer's disease
AMPA α-amino-3-hydroxy-5-methyl-4-isoxazolepropionic acid
ARC arcuate nucleus
α-MSH α-melanocyte-stimulating hormone
BAC bacterial artificial chromosome
Bv *Bombina variegata*
Ca^{2+} calcium ions
cAMP cyclic adenosine monophosphate
CCI constriction injury model
CFA complete Freund's adjuvant
DRG dorsal root ganglia
EG-VEGF endocrine gland-derived vascular endothelial growth factor
EGFP enhanced green fluorescent protein
EMT epicardial–mesenchymal transition
EPDC epicardial-derived progenitor cells
ERK extracellular signal-regulated protein kinases
FSH follicle-stimulating hormone
GnRH gonadotropin-releasing hormone
GnRHR gonadotrophin-releasing hormone receptors
GPCR G protein-coupled receptor
hCG human chorionic gonadotropin
hPROKR1 human prokineticin receptor 1
hPROKR2 human prokineticin receptor 2
HPG hypothalamic–pituitary–gonadal
ICV intracerebroventricular
IHH idiopathic hypogonadotropic hypogonadism

IUGR	intrauterine growth restriction
KS	Kallmann syndrome
LH	luteinizing hormone
MAPK	mitogen-activated protein kinase
MIT-1	mamba intestinal toxin 1
MRAP2	melanocortin receptor accessory protein 2
mProkr1	mouse prokineticin receptor 1
mProkr2	mouse prokineticin receptor 2
NGN1	neurogenin 1
nIHH	normosmic IHH
NPY	neuropeptide Y
OB	olfactory bulb
OV	olfactory ventricle
PAG	periaqueductal gray
PD	Parkinson's disease
PKCε	protein kinase Cε
PPARγ	peroxisome proliferator-activated receptor gamma
PROKR1	prokineticin receptor 1
PROKR2	prokineticin receptor 2
PROKs	prokineticins
RMS	rostral migratory stream
RVM	rostro ventromedial medulla
SCN	suprachiasmatic nucleus
SFO	subfornical organ
SNI	spared nerve injury
SVZ	subventricular zone
TRPV1	transient receptor potential vanilloid 1
TSHZ1	teashirt zinc finger family member 1
VEGF	vascular endothelial growth factor

1. Discovery of prokineticins and their receptors

A prokineticin-like peptide was first reported in 1980, when Joubert and Strydom isolated venom protein A from the nontoxic constituents in the venom of black mamba snake (*Dendroaspis polylepis*).[1] Later, Schweitz et al. demonstrated that this protein can induce the contraction of the ileum in guinea pig and called it MIT-1 (mamba intestinal toxin 1).[2,3]

A protein with similar size (~8 kDa) was isolated from the skin secretion of the fire-bellied toad (*Bombina variegata*) and demonstrated to also cause potent contractions of gastrointestinal smooth muscles.[4] This protein was named Bv8 to indicate its species of origin (*B. variegata*) and its molecular mass of 8 kDa.[4] Bv8 and MIT-1 share 58% identity in amino acid sequence.[4]

In the search for the mammalian homologs of MIT-1 and Bv8, Zhou and colleagues isolated and characterized two human cDNAs and named the peptides encoded by these cDNAs as "prokineticins" to reflect the potent actions on gastrointestinal contractility.[5] One cDNA encodes a protein of 86 amino acids (Prokineticin1; PROK1) with a high degree of homology with MIT1; and the other encodes an 81-amino acid protein (Prokineticin2; PROK2) highly related to Bv8.[5]

In the same year, Ferrara and colleagues identified a protein specifically expressed in the endocrine tissues (placenta, ovary, testis, and adrenal gland), acting as a mitogenic agent for endothelial cells from these tissues.[6] Based on its similar biological activity to vascular endothelial growth factor (VEGF), this protein was called endocrine gland-derived vascular endothelial growth factor (EG-VEGF).[6] However, except for the similarities in the biological actions, there is no correlation between VEGF and EG-VEGF. Sequence analysis indicates that EG-VEGF and PROK1 are the same molecule.[5,6] In this chapter, we will use the designations of PROK1 and PROK2 to maintain consistency with the nomenclature in the literature.

Zhou and colleagues initially found that binding of PROK1 to the membranes prepared from guinea pig ileum can be displaced by GTPγS, suggesting that the prokineticin receptor may belong to the family of G protein-coupled receptors (GPCRs).[5] Soon after the discovery of prokineticins, three independent research groups identified two closely related GPCRs as the receptors for prokineticins.[7–9] These receptors were named prokineticin receptor 1 (PROKR1) and prokineticin receptor 2 (PROKR2), respectively. PROKR1 and PROKR2 belong to the neuropeptide Y (NPY) receptor class, sharing 85% amino acid identity, with the most divergent part residing in the N-terminal extracellular domains.[7–9]

2. Genomic structures of prokineticins

The *PROK1* gene maps to the regions of human chromosome 1p13.1 and mouse chromosome 3, respectively.[10] The gene organization is highly conserved and composed of three exons with no alternative splicing product. The first exon encodes 19 residues corresponding to the signal peptide and the first 5 amino acids (AVITG) of the mature protein. The second exon encodes 42 amino acids with 6 cysteine residues, and the third exon encodes 39 amino acids with 4 cysteine residues.[11,12] (Fig. 1).

The *PROK2* gene maps to the regions of human chromosome 3p21 and mouse chromosome 6, respectively.[13,14] There are four exons for *PROK2*.

```
                Signal peptide
mPROK1  MRGAV--------HIFIMLLLATASDCAVITGACERDIQCGAGTCCAISLWLRGLRLCTP
hPROK1  MRGAT--------RVSIMLLLVTVSDCAVITGACERDVQCGAGTCCAISLWLRGLRMCTP
mPROK2  MGDPRCAPLLLLLLL-PLLFTPPAGDAAVITGACDKDSQCGGGMCCAVSIWVKSIRICTP
hPROK2  MRSLCCAPLLLLLLLPPLLLTPRAGDAAVITGACDKDSQCGGGMCCAVSIWVKSIRICTP
mPROK2L MGDPRCAPLLLLLLL-PLLFTPPAGDAAVITGACDKDSQCGGGMCCAVSIWVKSIRICTP
hPROK2L MRSLCCAPLLLLLLLPPLLLTPRAGDAAVITGACDKDSQCGGGMCCAVSIWVKSIRICTP
                                        *     *    **           *

mPROK1  LGREGEECHPGSHK---------------------IPFLRKRQHHTCPCSPSLLCSRFPD
hPROK1  LGREGEECHPGSHK---------------------VPFFRKRKHHTCPCLPNLLCSRFPD
mPROK2  MGQVGDSCHPLTRK---------------------VPFWGRRMHHTCPCLPGLACLRTSF
hPROK2  MGKLGDSCHPLTRK---------------------VPFFGRRMHHTCPCLPGLACLRTSF
mPROK2L MGQVGDSCHPLTRKSHVANGRQERRRAKRRKRKKEVPFWGRRMHHTCPCLPGLACLRTSF
hPROK2L MGKLGDSCHPLTRKNNFGNGRQERRKRKRSKRKKEVPFFGRRMHHTCPCLPGLACLRTSF
               *     ▲                                 * *    *

mPROK1  GRYRCFRDLKNANF   105
hPROK1  GRYRCSMDLKNINF   105
mPROK2  NRFICLARK-----   107
hPROK2  NRFICLAQK-----   108
mPROK2L NRFICLARK-----   128
hPROK2L NRFICLAQK-----   129
            *
```

Fig. 1 Amino acid sequences alignment of mouse PROK1 (mPROK1, NM_001044382.2, NP_001037847.1), human PROK1 (hPROK1, NM_032414.2, NP_115790.1), mouse PROK2 (mPROK2, NM_001037539.2, NP_001032628.1), human PROK2 (hPROK2, NM_021935.3, NP_068754.1), long form of the mouse PROK2 (mPROK2L, NM_015768.2, NP_056583.1), and long form of the human PROK2 (hPROK2L, NM_001126128.1, NP_001119600.1). The N-terminal AVITGA sequences are underlined. Ten conservative cysteine residues are marked (*). Arrows indicate the splicing sites for exons. Triangle indicates the proteolytic cleavage site for PROK2β.

Exon 1 encodes the signal peptide and the first five amino acids (AVITG) of the mature protein. Exon 2 encodes 42 amino acids with 6 cysteine residues. Exon 3 encodes the 21 amino acid insert that is present in an alternative splicing product. Exon 4 encodes 34 amino acids with 4 cysteine residues. The third exon of *PROK2* is subjected to alterative splicing, resulting in two mature proteins.[15] PROK2 is encoded by exons 1, 2, and 4; whereas PROK2L, a long form of the PROK2 with 21 additional amino acids, is encoded by all 4 exons of *PROK2*[15] (Fig. 1).

The human and mouse *PROK2* promoter sequences are highly conserved.[12,14] The *PROK2* promoter region has several *E*-boxes (CACGTG) that are recognized by the members of the basic helix–loop–helix family such as CLOCK and BMAL1[16] (implicated in the role of PROK2 in circadian regulation in the suprachiasmatic nucleus—SCN) and NGN1 and MASH1 (implicated in the role of PROK2 in the olfactory bulb—OB).[17]

3. Protein structures of prokineticins

The mature PROK1 protein is composed of 86 amino acids. The precursor is a protein of 105 amino acids, including a signal peptide of 19 amino acids.[5] The mature PROK2 is composed of 81 amino acids, while the long form PROK2L contains 102 amino acids.[5,15] The secreted PROK2L peptide is processed into a smaller peptide of 47 amino acids (PROK2β) by proteolytic cleavage.[15] Functional study indicated that PROK2β is a selective ligand for PROKR1[15] (Fig. 1).

Human PROK1 and PROK2 share 44% identity.[18] PROK1 and PROK2 have a structurally conserved motif characterized by an N-terminal AVITGA sequences and a carboxyl-terminal cysteine-rich domain that forms five disulfide bridges with conserved spacing.[19] Substitutions, deletions, and insertions to the conserved N-terminal hexapeptides result in the loss of agonist activity. Substitution of the first N-terminal alanine with methionine or the addition of a methionine to the N-terminus results in prokineticins with antagonist activity.[19,20] Furthermore, mutations in select cysteine residues in the C-terminal domain also result in prokineticins without biological activity.[19]

PROK1 and PROK2 fold into a polarized ellipsoid structure with one side containing a net positive charge and the opposite side with hydrophobic residues.[21] The C- and N-termini are exposed on the surface, whereas the more charged residues are buried inside the molecule.[21] The high degree of disulfide cross-linking gives rise to a remarkably stable compact protein that is highly resistant to protease degradation.[19,21]

4. Genomic and protein structures of prokineticin receptors

The *PROKR1* gene is located in human chromosome 2p13.3 and mouse chromosome 6, respectively. The *PROKR2* gene is located in human chromosome 20p13 and mouse chromosome 2, respectively. The coding regions of *PROKR1* and *PROKR2* genes comprise two exons, which is separated at the border of transmembrane domain III with the highly conserved DRY motif. It should be noted that another exon is located in the 5′-UTR of h*PROKR1*, m*Prokr1* and m*Prokr2*, but not h*PROKR2*.

The prokineticin receptors consist of an extracellular N-termini, seven transmembrane helices, three intracellular loops, three extracellular loops, and cytoplasmic C-terminal tails. PROKR1 and PROKR2 are similar to the NPY receptor with two Cys residues located on the first and second extracellular loop giving rise to a highly conserved disulfide bond in family A GPCRs.[7–9]

PROKR1 is composed of 393 amino acids, whereas PROKR2 is composed of 384 amino acids. PROK1 and PROK2 can bind and activate both receptors. However, the signal transduction efficacy of PROKR1 is slightly higher than that of PROKR2. Site-directed mutagenesis demonstrated that replacement of a basic amino acid (Lys-175) of PROKR1 with a neutral amino acid in PROKR2 (Asn-166) may underlie this difference in signaling.[22]

It has been shown that activation of PROKRs leads to accumulation of inositol phosphate and mobilization of intracellular Ca^{2+} via Gq/11 proteins.[7–9] In addition, PROKRs may stimulate or inhibit cAMP accumulation through Gs or Gi proteins, respectively.[11,15] Furthermore, PROKRs have been shown to stimulate mitogen-activated protein kinase (MAPK) via Go protein-mediated signaling, released βγ subunits of G protein and phospholipase Cβ.[23]

5. Expression of prokineticins and their receptors

PROK1 and *PROK2* are coexpressed in various tissues, including brain, ovary, testis, placenta, adrenal cortex, peripheral blood cells, intestinal tract, heart, and bone marrow, but there are also some striking differences in tissue expression patterns of these two receptors.[24–26] A major difference is that, in humans, *PROK1* is predominantly expressed in steroidogenic organs, including ovary, testis, adrenal cortex, and placenta; whereas *PROK2* is primarily (but not exclusively) expressed in the central nervous system and nonsteroidogenic cells of the testes.[6,26–28]

The expression of prokineticins in some organs/tissues is also dynamic throughout physiological processes, such as circadian rhythm, menstrual cycle, and pregnancy. For instance, *PROK1* in the ovaries is regulated by human chorionic gonadotropin (hCG), estrogen, and follicle-stimulating hormone (FSH). And *PROK1* in the placenta is regulated by hCG, PPARγ, and hypoxia.[29] In the suprachiasmatic nuclei (SCN) of the anterior hypothalamus, the pacemaker for mammalian circadian rhythms, *PROK2* is highly expressed in a circadian manner, which is regulated by the endogenous clockwork and other signaling such as vasopressin-V1a signaling.[27,30]

Although *PROKR1* and *PROKR2* are coexpressed in certain tissues, *PROKR1* is mainly expressed in peripheral tissues, including endocrine glands and organs of the reproductive system, the gastrointestinal tract, spleen, pancreas, lungs, heart, and blood cells.[7,8,26,31–35] *PROKR2* is abundantly expressed in the brain and testes.[26] Cheng et al. conducted a comprehensive in situ hybridization study on the expression of prokineticins and their receptor in adult mouse brain.[33] *Prok1* is expressed exclusively in the brainstem; however, *Prok2* is expressed in the SCN, islands of Calleja and medial preoptic area, OB, nucleus accumbens shell, hypothalamic arcuate nucleus (ARC), and amygdala. *Prokr2* is widely expressed in the brain, including olfactory regions, cortex, thalamus and hypothalamus, septum and hippocampus, habenula, amygdala, nucleus tractus solitarius, and circumventricular organs such as subfornical organ (SFO), median eminence, area postrema, mammillary nuclei, periaqueductal gray (PAG), and dorsal raphe. In contrast, *Prokr1* is found only in a few brain regions, including the olfactory regions, dentate gyrus, zona incerta, and dorsal motor vagal nucleus.[33]

6. Physiological function of prokineticins and their receptors

The first physiological function of prokineticins signaling identified is to promote the gastrointestinal contraction.[1–5] However, further studies have demonstrated that prokineticins signaling is involved in a myriad of physiological functions, including circadian rhythms, neurogenesis, ingestive behaviors, nociception, and mood regulation.

6.1 Circadian rhythms regulation

The physiology and behaviors of almost all organisms exhibit circadian (~24 h) rhythms to adapt to the environmental changes imposed by the daily revolutions of the earth. The mammalian pacemaker for circadian rhythms resides in the SCN of the anterior hypothalamus, which coordinates daily cycles of behavior and physiology.[36–40] The molecular mechanisms of this endogenous clock in the SCN, and cells throughout the body, have been elucidated as consisting of several autoregulatory transcriptional/translational feedback loops.[38,39,41–44]

In an attempt to study the function of prokineticins in the central nervous system, Zhou and colleagues systematically mapped the mRNA distribution of prokineticins and their receptors in the adult mouse brain.[33] Impressively, *Prok2* mRNA is highly expressed in the SCN in a circadian fashion, with high levels during the day and low or undetectable levels at night. They further showed that *Prok2* oscillation is driven by the endogenous circadian clockwork, as *Prok2* mRNA expression in the SCN is completely absent or blunted in mutant mice lacking functional clockwork, such as *Clock* mutant and $Cry1^{-/-}$ $Cry2^{-/-}$ mice.[27] In vitro transcription assay indicated that the positive elements of the clockwork, Clock and Bmal1, bind to the *E*-boxes (CACGTG) residing in the *Prok2* proximal promoter and activate the transcription of *Prok2*.[27]

In situ hybridization also indicated that *Prokr2* is expressed in primary SCN output target areas, including the ventral lateral septum, medial preoptic area, subparaventricular zone, paraventricular nucleus, dorsomedial hypothalamic nucleus, lateral hypothalamic area, and paraventricular thalamic nucleus.[16] The projection of SCN-derived PROK2 to these areas is further validated by a study using a bacterial artificial chromosome (BAC) transgenic mouse line, in which the enhanced green fluorescent protein (EGFP) reporter gene expression was driven by the *Prok2* promoter.[45] Intracerebroventricular (ICV) delivery of PROK2 into the lateral ventricle during subjective night, when endogenous *Prok2* was low, inhibits the nocturnal wheel-running activity of rats,[16] further supporting a role of PROK2 in the circadian control of locomotion.

The importance of PROK2/PROKR2 signaling in the circadian regulation was also demonstrated in mice deficient in *Prok2* or *Prokr2*, respectively. The $Prok2^{-/-}$ mice show attenuated circadian rhythmicity in a variety of behaviors and physiology tests, including locomotor activity, sleep/wake, body temperature, hormone, as well as peripheral clock gene expression.[13,46] The $Prokr2^{-/-}$ mice show a similar attenuation in their daily rhythms.[47] Nevertheless, clock gene expression is not altered in the SCN of either $Prok2^{-/-}$ or $Prokr2^{-/-}$ mice, consistent with the supposed role of PROK2 as an output molecule.[13,47] Recently, Qu et al. indicated that zinc finger protein ZBTB20 may enhance P300-mediated *Prokr2* transcription. Depletion of *Zbtb20* in the nervous system results in similar circadian phenotype of $Prok2^{-/-}$ and $Prokr2^{-/-}$ mice.[48]

By using a zebrafish model, Chen et al. demonstrated that PROK2 affects sleep and wake behavior in a light-dependent but circadian-independent manner.[49] In light, *prok2* overexpression increases sleep partly through inducing the expression of galanin, a sleep-inducing peptide.[49]

A novel role of PROK2–PROKR2 in circadian regulation was recently proposed. Zhou et al. identified divergent expression of PROKR2 in the nocturnal mouse and the diurnal monkey in the retinorecipient SCN and the superior colliculus, projection targets of the intrinsically photosensitive retinal ganglion cells.[50] They further showed that PROK2 signaling may inhibit the arousal levels in nocturnal mouse, but stimulate the arousal levels in the diurnal monkey.[50] They thus proposed that either the mammalian diurnality or nocturnality may be determined by the differential signaling of PROK2 from the intrinsically photosensitive retinal ganglion cells onto their retinorecipient brain targets.[50]

6.2 Ingestive behaviors

The role of prokineticin system in the ingestive behaviors was first studied by Negri et al. in 2004. They found that ICV injection of Bv8 suppressed feeding, but stimulated drinking in rats.[51] They further identified a brain area-specific effect for Bv8 on feeding and drinking.[51] Delivery of Bv8 into the ARC selectively suppressed feeding without affecting drinking, whereas injection of Bv8 into the SFO stimulated drinking but did not affect feeding. Bv8 injections into other brain areas left rat ingestive behaviors unchanged.[51] Indeed, Ferguson and colleagues demonstrate that PROK2 modulates the excitability of SFO, through the activation of a MAP kinase cascade, which in turn modulates Na^+ and K^+ conductance.[52,53]

Gardiner et al. further demonstrated that ICV administration of PROK2 increases *c-Fos* expression in proopiomelanocortin neurons of the ARC. The anorectic effect of ICV-delivered PROK2 seems to be mediated by α-melanocyte-stimulating hormone (α-MSH), as PROK2 increases the release of α-MSH from ex vivo hypothalamic explants and ICV coadministration of the α-MSH antagonist Agouti-related peptide blocks the anorexigenic effects of PROK2.[54]

Interestingly, Chaly et al. recently showed that inhibition of PROKR signaling may mediate the stimulatory effect of melanocortin receptor accessory protein 2 (MRAP2) on food intake. They showed that MRAP2 specifically inhibits PROKR1 signaling, and *Mrap2*$^{-/-}$ mice are hypersensitive to PROKR1 stimulation.[55]

Peripheral administration of PROK2 also potently reduces food intake. Beale et al. demonstrated that intraperitoneal delivery of PROK2 significantly increases *c-Fos* expression in the dorsal motor vagal nucleus of the brainstem.[56] This effect is mediated by PROKR1, instead of PROKR2, as the anorectic effect of PROK2 is intact in *Prokr2*$^{-/-}$ mice but abolished in *Prokr1*$^{-/-}$ mice and in wild-type (WT) mice treated with a PROKR1 antagonist.[56]

6.3 Mood regulation and stress response

As disrupted circadian rhythms are strictly associated with mood disorders, we sought to analyze the effect of PROK2 on mood regulation after showing its roles in circadian regulation. ICV infusion of PROK2 increases anxiety- and depression-like behaviors. Conversely, *Prok2*$^{-/-}$ mice show reduced anxiety- and depression-like behaviors. These mice also show impaired responses to new environments, in terms of locomotor activity, arousal, body temperature, and food intake.[57] We further showed that fasting-induced arousal is absent in *Prok2*$^{-/-}$ mice and *Prok2*$^{-/-}$ mice enters torpor after fasting.[58] Similarly, *Prokr2*$^{-/-}$ mice show spontaneous torpor, which is exaggerated by fasting.[59]

The role of prokineticin system in mood regulation was also studied in human populations. Kishi et al. suggest a possible association of *PROKR2* with mood disorders in the Japanese population.[60] Further, they also detected a significant association between *PROKR2* and methamphetamine dependence in allele/genotype-wise and haplotype-wise analysis.[61] However, due to the small sample size, these data need to be confirmed in large population and in other ethnicities.

6.4 Nociception regulation

The first evidence of a pronociceptive role of prokineticins was derived from the observation that systemic injection of Bv8/PROK2 in rodents induces hyperalgesia to mechanical and thermal stimuli.[4,62] *Prok2*$^{-/-}$ mice displayed strong reduction in nociception induced by thermal and chemical stimuli, indicating an important role for endogenous PROK2 in the pain sensitization.[63]

Both PROKR1 and PROKR2 are expressed in the superficial layers of the spinal cord, dorsal root ganglia (DRG), and peripheral terminal of nociceptors. *Prokr1*$^{-/-}$ mice show impaired responsiveness to noxious heat, mechanical stimuli, capsaicin, and protons.[64] Bv8 causes hyperalgesia in WT mice and sensitizes the actions of capsaicin, which is impaired in *Prokr1*$^{-/-}$ mice.[64] Furthermore, the number of neurons that responded with intracellular calcium increase to Bv8 is five times lower in *Prokr1*$^{-/-}$ DRG cultures than in WT cultures.[64]

Mechanistically, Negri et al. showed that the majority of PROKR1-positive DRG neurons also express transient receptor potential vanilloid 1 (TRPV1), a critical ion channel in the nociceptive regulation.[65,66] Bv8 enhances the inward current of TRPV1 via a pathway involving activation of protein kinase Cε (PKCε).[65] In addition, Hu and colleagues also proposed some additional mechanism regarding PROK2 in pain regulation. They demonstrated that PROK2 can downregulate the function of the $GABA_A$ receptor and increases the activity of acid-sensing ion channels and P2X receptor in primary sensory neurons.[67–69]

It was also suggested that prokineticins are able to modulate central pain mechanism.[70] Maione and colleagues demonstrated that microinjection of Bv8 into the PAG exerts a pronociceptive effect by increasing the intrinsic GABAergic tone which is responsible for the inhibition of PAG antinociceptive output neurons impinging on rostro ventromedial medulla (RVM) neurons.[70]

6.4.1 Inflammation pain

Negri et al. found that *Prokr1*$^{-/-}$ mice show reduced pain-related behaviors after acute inflammation elicited by mustard oil or after chronic inflammation.[64] And both *Prokr1*- and *Prokr2*-null mice show reduced pain-related hypersensitivity produced by complete Freund's adjuvant (CFA), suggesting a role of prokineticin system in the inflammation pain.[71]

They further demonstrated that in a CFA-induced animal model of chronic inflammation, inflammation is highly correlated with an overexpression of PROK2 in the granulocytes that infiltrate the inflamed tissue and this upregulation is responsible for inflammation-associated hyperalgesia.[71,72] Importantly, the hyperalgesia induced by CFA is abolished by a PROKR1 antagonist, implying a therapeutic potential for PROKRs antagonists in inflammation pain.[71] Indeed, Kurosaka et al. demonstrated that chronic treatment with prokineticin receptor antagonists significantly reduces thermal hypersensitivity and the histological damage in a mouse model of type II collagen-induced arthritis.[73,74]

6.4.2 Neuropathic pain

Neuropathic pain is hyperalgesia and allodynia caused by damage or disease affecting the somatosensory nervous system. The involvement of prokineticins system in the neuropathic pain was investigated in a variety of animal models, including chronic constriction injury model (CCI) and spared nerve injury (SNI), diabetes, and cancer-induced neuropathic pain.

CCI and SNI in mice induce overexpression of PROK2 and its receptor PROKR2 in peripheral nerve and the spinal cord. Treatment with prokineticin antagonists PC1 is effective in controlling and preventing neuropathic pain.[75,76] PC1 is also efficacious in controlling pain and immune system dysregulation in a mouse model of streptozotocin-induced diabetes.[77] And in a rat cancer-induced bond pain model developed by injecting tumor cells in the medullary cavity of rat tibia, mechanical hyperalgesia developed in parallel with increased levels of PROK2 in the spinal cord. Intrathecal delivery of a PROK2-neutralizing antibody significantly attenuates the pain-related behaviors.[78]

6.5 Neurogenesis

In mammals, neurogenesis occurs mainly during embryonic to early postnatal stages. However, there are certain areas in the adult mammalian brain (such as the OB and dentate gyrus of the hippocampus) that generates new neurons throughout life.[79–81] Our in situ hybridization data indicated that the *Prok2* expression pattern is complementary to that of *Prokr2*.[82] *Prok2* is expressed in the granular and periglomerular layers of the OB, whereas *Prokr2* is expressed in the areas active in neurogenesis during adulthood, such as subventricular zone (SVZ), the entire rostral migratory stream (RMS), and the ependyma and subependymal layers of the olfactory ventricle (OV). Furthermore, in vitro experiments show that PROK2 functions as a chemoattractant for SVZ-derived neuronal progenitors. *Prok2*$^{-/-}$ mice have markedly reduced OB, with loss of normal OB architecture, and the accumulation of neuronal progenitors in the RMS.[82] Matsumoto et al. also showed that *Prokr2*$^{-/-}$ mice, but not *Prokr1*$^{-/-}$ mice, show hypoplasia and abnormal architecture of OB.[83] These data thus indicate an essential role for PROK2–PROKR2 signaling in OB neurogenesis.

In the OB, the expression of *Prok2* is regulated by basic helix–loop–helix transcriptional factors Ngn1 and MASH1, by binding to the *E*-boxes in its promoter.[17] The expression of *Prokr2* in the OB may be regulated by zinc finger homeodomain factor teashirt zinc finger family member 1 (TSHZ1).[84] Indeed, *Tshz1* deletion in mice results in OB hypoplasia and severe olfactory deficits.[84]

6.6 Heart development

Nebigil et al. systematically studied the role of prokineticin system in heart development, which was summarized in a recent review.[85] They first showed the expression of *Prok2* and *Prokr1* in the heart and cardiac cells. PROK2/PROKR1 activation induces vessel-like formation in cultured cardiac endothelial cells. Overexpression of PROKR1 in the cardiomyocytes rescues the myocardium against myocardial infarction.[86] They further demonstrated that PROKR1 signaling in cardiomyocyte upregulates PROK2, which acts as a paracrine factor to triggering the proliferation/differentiation of epicardial-derived progenitor cells (EPDC). And PROKR1 plays a role in regulating the epicardial–mesenchymal transition (EMT) for EPDC formation. Genetic ablation of PROKR1 in epicardium leads to defective heart development,[87] whereas nonpeptide agonists for PROKR1 have cardioprotective effects.[88]

6.7 Role of prokineticin system in reproduction

6.7.1 Function in male reproductive system

Prok1 and *Prok2* genes are highly expressed mammalians testis.[89,90] PROK1 protein is mainly present in Leydig cells,[89,90] while PROK2 is present within seminiferous tubules in primary spermatocytes.[89] As an angiogenic mitogen, the expression of *Prok1* in Leydig cells suggests its role in angiogenesis during the early endocrine development of testis and in the adult testis as well as in Leydig cell tumor growth.[91]

6.7.2 Function in female reproductive system

Prok2 is undetectable in female reproductive organs[28,92–94]; whereas *Prok1* expression is abundant in ovarian stroma, especially in hilar region cells that produce steroids.[28] The expression of *Prok1* fluctuates according to the ovarian cycle, with significant expression in the granulosa of primordial primary and secondary cells, followed by a decline in the tertiary follicle.[93] Upon ovulation, *Prok1* mRNA increases in the corpus luteum.[93] It should be noted that the abnormal expression of *Prok1* gene is observed in several ovarian diseases, such as polycystic ovarian syndrome and hyperstimulation ovarian syndrome.[6,93,95]

6.7.3 Function in pregnancy

PROK1 is the only prokineticin expressed in the placenta and plays a key role in the success of pregnancy. The circulating levels of PROK1 are significantly increased during pregnancy,[92] and PROK1 signaling may regulate the expression of a host of implantation-related genes.[96] Evans et al. demonstrated that PROK1 enhances adhesion of trophoblast cells to fibronectin and laminin matrices and thus mediates fetal-maternal dialogue during implantation, by inducing expression of leukemia inhibitory factor.[97]

During pregnancy, the prokineticin system has been established as an important actor of placental development.[92] Peak *PROK1/PROKR1* expression occurs during the first trimester of pregnancy.[94] During this period, PROK1 controls extravillous trophoblastic cell migration, invasion, and formation of pseudovascular networks.[98] PROK1 also acts on fetal endothelial cells within the stroma and increases their proliferation, migration, invasion, and branching.[99]

The increase in *PROK1* expression in the myometrium during labor, and the presence of prokineticins and their receptors in the intrauterine tissues suggest a role of this system in the labor process.[100,101] In the myometrium, prokineticins induce proinflammatory cascades and increase contractility of smooth muscle myometrium cells, two processes involved in the onset of preterm and term labor.[101] In the uteroplacental unit, PROK1 and its receptors have been shown to confer intrauterine quiescence during the last trimester of pregnancy.[100]

Prokineticins have been associated with early pregnancy pathologies such as ectopic pregnancies[102] as well as repeated miscarriages.[103–105] Recently, mutations in *PROK1*, *PROKR1*, and *PROKR2* genes have been reported to be associated with repeated miscarriages,[106] while some variants seem to exhibit protective roles in the early stages of pregnancy.[105] Furthermore, PROK1 is increased in preeclampsia and IUGR (intrauterine growth restriction).[98,107] However, it is still unknown whether the increase in PROK1 levels is a cause or a consequence of these diseases.

7. Prokineticin system in diseases

7.1 Angiogenesis and cancer

Angiogenesis is a process through which new blood vessels form from preexisting vessels. In addition to its physiological role in development, angiogenesis is also a fundamental step in the transition of tumors from a benign state to a malignant one.[108,109] VEGF has been demonstrated to be a major contributor to angiogenesis. PROK1 was identified by Ferrara and colleagues as the first tissue-specific angiogenic factor, because in vivo delivery of PROK1 to the ovary elicits potent angiogenesis response, but PROK1 administration in the skin or skeletal muscle elicits no response.[6,10]

Increased numbers of $CD11b^{+}Gr1^{+}$ cells, which consist primarily of neutrophils, have been reported in tumor-bearing mice and in cancer patients[110–113]; these cells infiltrate tumors and stimulate angiogenesis and tumor growth.[114,115] Shojaei et al. reported that resistance to anti-VEGF antibody treatment in some xenograft tumors is correlated with tumor infiltration by $CD11b^{+}Gr1^{+}$ myeloid cells.[116] Subsequent studies revealed that PROK2 is strongly upregulated in $CD11b^{+}Gr1^{+}$ cells associated with these resistant tumors.[117,118] Blocking PROK2 signaling with a neutralizing antibody inhibits tumor angiogenesis and growth.[118] They also showed that G-CSF, produced by a subset of tumors, is a key inducer of PROK2 expression in myeloid cells.[118] Therefore, PROK2 produced by neutrophils is one of the mediators of

VEGF-independent angiogenesis in tumors. Targeting PROK2 provides a promising therapeutic approach to inhibit tumor growth.

7.2 Ischemia and neurodegeneration

Prokineticins are also involved in neuropathology such as ischemia and neurodegeneration. In a stroke model produced by occlusion of middle cerebral artery, Cheng et al. demonstrated that *Prok2* mRNA is induced in the ischemic cortex and striatum. Central delivery of PROK2 worsens infarct volume, whereas PROKR2 antagonist decreases infarct volume and improves behavioral outcome. In vitro experiments indicated that PROK2 is upregulated by several pathological stressors, including hypoxia, reactive oxygen species, and excitotoxic glutamate.[119] These data identify PROK2 as a deleterious mediator for cerebral ischemia and suggest antagonists of PROKRs as potential protective agents for stroke. Nevertheless, a recent in vitro study demonstrates a protective role of PROK2 in cerebral ischemia and ischemic tolerance.[120] Further studies are needed to address this discrepancy.

Recently, Gordon et al. identified a paradigm for compensatory neuroprotective PROK2 signaling in nigral dopaminergic neurons.[121] They first demonstrated that *Prok2* expression is highly induced in nigral dopaminergic neurons during early stages of degeneration in multiple models of Parkinson's disease (PD). They further showed that *PROK2* expression is increased in surviving nigral dopaminergic neurons from PD patients' brains. Functional studies demonstrate that PROK2 promotes mitochondrial biogenesis and activates ERK and Akt survival signaling pathways, thereby driving neuroprotection.[121]

A role of prokineticin system in the amyloid β (Aβ)-induced neuronal damage is proposed. *Prok2* and its receptors *Prokr1* and *Prokr2* mRNA are upregulated by Aβ. A PROKR antagonist reduces the neurotoxicity and normalizes the AMPA current induced by Aβ (1–42) in cortical cultures. Moreover, the antagonist completely rescues long-term potential impairment in hippocampal slices from an Alzheimer's disease (AD) mouse model. These results indicate that PROK2 antagonists may represent a new therapeutic approach for AD.[122,123]

7.3 Idiopathic hypogonadotropic hypogonadism (IHH)

IHH is a rare disorder characterized by inappropriate low serum concentrations of the gonadotropins, luteinizing hormone (LH), and follicle-stimulating hormone (FSH), in the presence of low circulating concentrations of sex steroids.[124–127] It is caused by deficient production, secretion, or action of gonadotropin-releasing hormone (GnRH), leading to impairment of sex-steroid synthesis and gametogenesis, and subsequent impaired sexual development and infertility.[125,128] IHH is clinically heterogeneous, frequently accompanied by nonreproductive abnormalities, such as anosmia/hyposmia, craniofacial abnormalities, deafness, digital anomalies, dental agenesis, renal agenesis, and neurological defects.[125,129–132] Approximately 60% of IHH patients are accompanied with an impaired sense of smell (anosmia/hyposmia), which is termed Kallmann syndrome (KS), whereas approximately 40% of affected individuals show normal sense of smell (normosmic IHH—nIHH).[126,130]

GnRH, a 10 amino acid polypeptide, is a key factor regulating the function of the hypothalamic–pituitary–gonadal (HPG) axis.[133,134] GnRH is synthesized in the hypothalamus, and then transported to the anterior pituitary gland through the pituitary portal vein blood system, which binds to gonadotrophin-releasing hormone receptors (GnRHR) on gonadotrophs to stimulate the synthesis and secretion of LH and FSH. LH and FSH enter the peripheral circulatory system and act on the gonads (testes and ovaries) to stimulate the production of steroid hormones (estrogen, progesterone, and testosterone) and the formation of gametes.[135,136]

GnRH neurons are located in the hypothalamus, but during development, GnRH neuron precursors are derived from the extracranial areas. At 6 weeks of human pregnancy, GnRH neurons first appear on the inside of the embryonic olfactory placode and then migrated along the axons of the olfactory vomeronasal neurons. At 6.5 weeks, GnRH neurons penetrate through the cribriform plate, enter the forebrain, and migrate to the developing OB. Next, the GnRH neurons migrate back along the submeningal region of the left and right hemisphere cracks, then migrate laterally, and finally reach the hypothalamus at approximately 14 weeks. Then it detaches from the axon of the olfactory neurons and go on to differentiate and mature. Therefore, there is an intimate relationship between OB development and GnRH neuron migration.[137–141]

The deficient OB development in the *Prok2*$^{-/-}$ mice prompted us to study the development of GnRH neurons and reproductive organs. As a result, we demonstrated that *Prok2*$^{-/-}$ mice have substantially reduced GnRH neurons in the hypothalamus. These mice have hypogonadotropic hypogonadism with underdeveloped gonads.[82] Therefore, *Prok2*$^{-/-}$ mice phenocopy the human KS. We further identified a homozygous *PROK2* mutation from a pedigree with KS. This *PROK2* mutation produced a truncated protein with no biological activity.[142]

Matsumoto et al. found that the OB and reproductive system (including testis, ovary, and uterus) of *Prokr2*$^{-/-}$ mice are hypoplasia and serum testosterone and FSH concentrations are decreased. Immunohistochemical analysis revealed the absence of GnRH neurons in the hypothalamus of *Prokr2*$^{-/-}$ mice, whereas *Prokr1*$^{-/-}$ mice are normal as WT mice.[83]

Based on the phenotype of *Prok2*- and *Prokr2*-null mice, Dode et al. identified 10 different mutations of *PROKR2* and 4 different mutations of *PROK2* in a cohort of 192 patients with KS using candidate gene strategy.[143] In the follow-up studies, mutations in *PROK2* and *PROKR2* have also been identified in nIHH patients.[144] Up to now, 10 *PROK2* (7 missense mutations, 2 frame-shift insertion/deletion mutations, and 1 splicing mutation) and 46 *PROKR2* mutations (41 missense mutations, 3 frame-shift mutations, and 2 nonsense mutation) (Fig. 2) have been identified in IHH patients.[143–161] It is reported that ~5% of IHH patients harbored *PROKR2* mutations.[151] In particular, the frequency of *PROKR2* mutations in KS patients is as high as 23.3% in the Maghreb region of North Africa.[162]

In vitro functional studies indicated that pathogenic mutations of *PROK2* and *PROKR2* impair the activity of ligand or receptor. IHH-associated *PROKR2* may affect signal transduction from four aspects: (1) mutation affects the amount of protein expressed; (2) mutation affects membrane localization of PROKR2; (3) mutation disrupts the binding of receptor PROKR2 to extracellular ligand PROK2; and (4) mutation impairs G protein-dependent signal transduction to varying degrees[144–148,150,163] (Table 1).

The majority of mutations in patients with IHH is heterozygous (*PROK2* 8/10, 80%; *PROKR2* 38/44, 79.3%).[143–161] However, only homozygous *Prokr2* or *Prok2* knockout mice show hypogonadism, whereas heterozygous *Prokr2* or *Prok2* mice are quite normal.[83] One possible explanation is the bigenic and oligogenic mutations; i.e., the patients carry

Fig. 2 The diagram of *PROKR2* (NM_144773.3, NP_658986.1) mutations identified in patients with idiopathic hypogonadotropic hypogonadism. Mutations reported in ClinVar database but not in literatures are shown in *red*.

Table 1 Functional consequences of IHH-associated *PROKR2* mutations.

PROKR2 mutation	Protein expression	Membrane targeting	Ligand binding	Activities				References
				$[Ca2+]_i$	Gs	pERK	β-Arrestin recruitment	
A51T	—	—	—	—	↓	↓	—	145
R80C	—	—	—	—	↓	—	×	145
R85C	—/↓	↓	↓	—/↓	—	↓	—	144,145,148,150
R85G	↓	↓	↓	↓	N/A	↓	N/A	149
R85H	↓	—	↓	↓	—	↓	—	145,148,150
Y113H	↓	N/A	N/A	↓	N/A	↓	N/A	144
V115 M	↓	N/A	N/A	↓	N/A	↓	N/A	144
V158I	↓	N/A	N/A	—	↓	N/A	N/A	146
R164Q	—/↓	—	—	↓	—	↓/×	—	144,145,147,148,150
L173R	—/↓	↓	↓	↓	↓	↓	N/A	144,146,147
W178S	↓	↓	×	×	N/A	↓/×	N/A	144,147,150
S188 L	↓	N/A	N/A	↓	N/A	↓	N/A	144
Q210R	↓	—	↓/×	↓/×	×	×	×	145,148,150
G234D	↓	↓	N/A	N/A	N/A	N/A	N/A	145,147
R248Q	—	N/A	N/A	↓	N/A	—	N/A	144

Continued

Table 1 Functional consequences of IHH-associated *PROKR2* mutations.—cont'd

PROKR2 mutation	Protein expression	Membrane targeting	Ligand binding	Activities				References
				$[Ca2+]_i$	Gs	pERK	β-Arrestin recruitment	
T260 M	↓	N/A	N/A	↓	↓	N/A	N/A	146
R268C	—/↓	—	↓	↓	↓	×	—	145,146,148,150
V274D	↓	N/A	N/A	↓	↓	N/A	N/A	146,164
P290S	—/↓	↓	×	×	N/A	N/A	N/A	147,148,150
M323I	↓	↓	—	—	—	↓	—	145,148,150
N325K	N/A	N/A	N/A	×	×	×	N/A	161
V331M	—/↓	—	↓	↓	—	—/↓	—	144–146,148,150
V334M	↑	N/A	N/A	↓	↓	N/A	N/A	146
R357W	—	—	—	↓	—	—	—	144,145

—, unchanged; ↓, decreased; ×, none; ↑, increased; N/A, data not available.

pathogenic mutations in more than one causative genes. In 2010, Sykiotis et al. demonstrated that 11.3% of IHH patients with detectable rare protein-altering variants carried more than one gene mutation by analyzing eight IHH-associated genes.[165] It is reasonable to speculate that this rate will be even higher when more IHH-associated genes are identified. Indeed, our recent genetics study in a Chinese IHH cohort showed that up to 26.4% patients carried bigenic and oligogenic mutations (Zhao et al., unpublished observation).

Another explanation is possible dominant-negative effect of mutant PROKR2 on the WT receptor. Cox et al. recently measured the signaling in cells cotransfected with WT and mutant PROKR2s. They found that three mutants (G234D, W178S, and N325K) exert dominant-negative effect, decreasing WT receptor signaling[161] (Fig. 2).

In 2014, Sbai et al. systematically studied the function of 21 IHH-associated *PROKR2* mutations on different signaling pathways (Gαs, Gαq, ERK1/2, and β-arrestin recruitment). They found that some mutant receptors show biased signaling. For instance, four mutant receptors (R85C, R85H, R164Q, and V331M) with defective Gq signaling can still recruit β-arrestins on ligand binding. In contrast, the R80C receptor can activate three types of G proteins but cannot recruit β-arrestins. R268C receptor can recruit β-arrestins and activate the Gq and Gs signaling pathways but cannot activate ERK1/2 pathway.[145] We recently also identified two novel IHH-associated *PROKR2* mutations that selectively disrupted Gq pathway, but kept the other pathways intact (Zhao et al., unpublished observation).

Most IHH-associated variants are private mutations (i.e., only occurred once in a cohort); however, some mutations occur frequently in IHH patients. Especially, L173R *PROKR2* was the most prevalent founder mutation, occurring in 2.4% of unrelated IHH patients of diverse ethnicity, including Mexicans, Caucasians, Brazilians, and Maghrebis. So far, L173R *PROKR2* is the oldest IHH-associated mutation which originated approximately 5500–12,000 years ago.[166] Why these founder mutations that cause impaired reproduction can survive in strict natural selection for thousands of years? Avbelj Stefanija et al. proposed that these mutations may have some selective advantages in the human evolutionary process that are conducive to survival in harsh environments. For example, it might help energy preservation during cold environment, or adapt to caloric restriction easily during famine.[166] Indeed, mice deficient in the signaling of PROK2–PROKR2 go to torpor during fasting, resulting in less body weight loss than WT controls.[58,59] Recently, we also identified a possible founder *PROKR2* (c.533G > C; p.W178S) mutation, which occurred in 10 out of 135 Chinese IHH patients (Zhao et al., unpublished observation).

Recent studies suggest overlapping phenotypes/genotypes between KS and congenital hypopituitarism, a rare condition that may be associated with complex midline defects of the forebrain, including holoprosencephaly and septo-optic dysplasia.[149,167] It is worth noting that *PROKR2* mutations were identified in patients with hypopituitarism.[149,167,168]

Functional hypothalamic amenorrhea is a reversible form of GnRH deficiency caused by stressors such as nutritional deficits (i.e., excessive weight loss), excessive exercise, or psychological distress.[169,170] In 2011, several loss-of-function mutations in IHH-associated genes were identified in women with hypothalamic amenorrhea, including two *PROKR2* mutations.[171] This indicates that *PROKR2* mutations may contribute to the variable susceptibility of women to the functional changes in GnRH secretion that characterizes hypothalamic amenorrhea.

8. Conclusion and perspective

During the past 20 years, the prokineticin system has been reported to function in a panel of physiological functions. And mutations or dysfunction of these factors has also been implicated in several disorders. Targeting prokineticin system may provide therapeutic approach for diseases such as cancer, inflammation, pain, and neuropathology.

It should be noted that many data are obtained from pharmacological experiments. However, as PROK1 and PROK2 can activate both receptors, verification of these data with respective knockout mice may further clarify the functional target.

Furthermore, most of the studies were conducted with systemic drug delivery or conventional knockout mice. It is thus unknown whether the findings are the direct consequence or secondary effect. For instance, hypothalamic GnRH deficiency may confound the local effect of prokineticins in reproductive system. Therefore, tissue-specific knockout may help to dissect the target organs/tissues.

Although in vitro experiments have been conducted to understand the signaling pathways of prokineticins in some functions, the signaling pathways in a specific function is still largely unknown. Since some IHH-associated *PROKR2* mutations show biased signaling, knock-in mice carrying these mutations may help to understand the specific pathways in respective physiological/pathological functions.

Acknowledgments

This project was financially supported by the National High-Tech R&D Program (2015AA020502), National Basic Research Program of China (2012CB517904), National Natural Science Foundation of China (31371187, 81770780, and 81728013), the Key Research and Development Programs from Hunan Province (2018DK2010, 2018DK2013), the Fundamental Research Funds for the Central Universities of Central South University (1053320181293 and 1053320181270), Natural Science Foundation of Hunan Province (2018JJ3573), and Training Program Foundation for Excellent Innovation Youth of Changsha.

We apologize to the authors whose work could not be cited in this review due to space limitations but appreciate all of the many contributions to the large body of literature we have tried to summarize here.

Conflict of interest

The authors declare no conflicts of interest.

References

1. Joubert FJ, Strydom DJ. Snake venom. The amino acid sequence of protein A from *Dendroaspis polylepis polylepis* (black mamba) venom. *Hoppe Seylers Z Physiol Chem*. 1980;361:1787–1794.
2. Schweitz H, Bidard JN, Lazdunski M. Purification and pharmacological characterization of peptide toxins from the black mamba (*Dendroaspis polylepis*) venom. *Toxicon*. 1990;28:847–856.
3. Schweitz H, Pacaud P, Diochot S, Moinier D, Lazdunski M. MIT(1), a black mamba toxin with a new and highly potent activity on intestinal contraction. *FEBS Lett*. 1999;461:183–188.
4. Mollay C, Wechselberger C, Mignogna G, et al. Bv8, a small protein from frog skin and its homologue from snake venom induce hyperalgesia in rats. *Eur J Pharmacol*. 1999;374:189–196.
5. Li M, Bullock CM, Knauer DJ, Ehlert FJ, Zhou QY. Identification of two prokineticin cDNAs: recombinant proteins potently contract gastrointestinal smooth muscle. *Mol Pharmacol*. 2001;59:692–698.
6. LeCouter J, Kowalski J, Foster J, et al. Identification of an angiogenic mitogen selective for endocrine gland endothelium. *Nature*. 2001;412:877–884.
7. Masuda Y, Takatsu Y, Terao Y, et al. Isolation and identification of EG-VEGF/prokineticins as cognate ligands for two orphan G-protein-coupled receptors. *Biochem Biophys Res Commun*. 2002;293:396–402.
8. Soga T, Matsumoto S, Oda T, et al. Molecular cloning and characterization of prokineticin receptors. *Biochim Biophys Acta*. 2002;1579:173–179.
9. Lin DC, Bullock CM, Ehlert FJ, Chen JL, Tian H, Zhou QY. Identification and molecular characterization of two closely related G protein-coupled receptors activated by prokineticins/endocrine gland vascular endothelial growth factor. *J Biol Chem*. 2002;277:19276–19280.
10. LeCouter J, Lin R, Frantz G, Zhang Z, Hillan K, Ferrara N. Mouse endocrine gland-derived vascular endothelial growth factor: a distinct expression pattern from its human ortholog suggests different roles as a regulator of organ-specific angiogenesis. *Endocrinology*. 2003;144:2606–2616.
11. Lin R, LeCouter J, Kowalski J, Ferrara N. Characterization of endocrine gland-derived vascular endothelial growth factor signaling in adrenal cortex capillary endothelial cells. *J Biol Chem*. 2002;277:8724–8729.

12. LeCouter J, Lin R, Tejada M, et al. The endocrine-gland-derived VEGF homologue Bv8 promotes angiogenesis in the testis: localization of Bv8 receptors to endothelial cells. *Proc Natl Acad Sci U S A*. 2003;100:2685–2690.
13. Li JD, Hu WP, Boehmer L, et al. Attenuated circadian rhythms in mice lacking the prokineticin 2 gene. *J Neurosci*. 2006;26:11615–11623.
14. Jilek A, Engel E, Beier D, Lepperdinger G. Murine Bv8 gene maps near a synteny breakpoint of mouse chromosome 6 and human 3p21. *Gene*. 2000;256:189–195.
15. Chen J, Kuei C, Sutton S, et al. Identification and pharmacological characterization of prokineticin 2 beta as a selective ligand for prokineticin receptor 1. *Mol Pharmacol*. 2005;67:2070–2076.
16. Cheng MY, Bullock CM, Li C, et al. Prokineticin 2 transmits the behavioural circadian rhythm of the suprachiasmatic nucleus. *Nature*. 2002;417:405–410.
17. Zhang C, Ng KL, Li JD, et al. Prokineticin 2 is a target gene of proneural basic helix-loop-helix factors for olfactory bulb neurogenesis. *J Biol Chem*. 2007;282:6917–6921.
18. Maldonado-Perez D, Evans J, Denison F, Millar RP, Jabbour HN. Potential roles of the prokineticins in reproduction. *Trends Endocrinol Metab*. 2007;18:66–72.
19. Bullock CM, Li JD, Zhou QY. Structural determinants required for the bioactivities of prokineticins and identification of prokineticin receptor antagonists. *Mol Pharmacol*. 2004;65:582–588.
20. Negri L, Lattanzi R, Giannini E, et al. Biological activities of Bv8 analogues. *Br J Pharmacol*. 2005;146:625–632.
21. Boisbouvier J, Albrand JP, Blackledge M, et al. A structural homologue of colipase in black mamba venom revealed by NMR floating disulphide bridge analysis. *J Mol Biol*. 1998;283:205–219.
22. Peng Z, Tang Y, Luo H, et al. Disease-causing mutation in PKR2 receptor reveals a critical role of positive charges in the second intracellular loop for G-protein coupling and receptor trafficking. *J Biol Chem*. 2011;286:16615–16622.
23. Yin W, Liu H, Peng Z, Chen D, Li J, Li JD. Mechanisms that underlie the internalization and extracellular signal regulated kinase 1/2 activation by PKR2 receptor. *Cell Signal*. 2014;26:1118–1124.
24. Ngan ES, Tam PK. Prokineticin-signaling pathway. *Int J Biochem Cell Biol*. 2008;40: 1679–1684.
25. Negri L, Lattanzi R, Giannini E, Canestrelli M, Nicotra A, Melchiorri P. Bv8/prokineticins and their receptors: a new pronociceptive system. *Int Rev Neurobiol*. 2009;85:145–157.
26. Negri L, Ferrara N. The prokineticins: neuromodulators and mediators of inflammation and myeloid cell-dependent angiogenesis. *Physiol Rev*. 2018;98:1055–1082.
27. Cheng MY, Bittman EL, Hattar S, Zhou QY. Regulation of prokineticin 2 expression by light and the circadian clock. *BMC Neurosci*. 2005;6:17.
28. Ferrara N, LeCouter J, Lin R, Peale F. EG-VEGF and Bv8: a novel family of tissue-restricted angiogenic factors. *Biochim Biophys Acta*. 2004;1654:69–78.
29. Traboulsi W, Brouillet S, Sergent F, et al. Prokineticins in central and peripheral control of human reproduction. *Horm Mol Biol Clin Invest*. 2015;24:73–81.
30. Li JD, Burton KJ, Zhang C, Hu SB, Zhou QY. Vasopressin receptor V1a regulates circadian rhythms of locomotor activity and expression of clock-controlled genes in the suprachiasmatic nuclei. *Am J Physiol Regul Integ Comp Physiol*. 2009;296: R824–R830.
31. Boulberdaa M, Urayama K, Nebigil CG. Prokineticin receptor 1 (PKR1) signalling in cardiovascular and kidney functions. *Cardiovasc Res*. 2011;92:191–198.
32. Attramadal H. Prokineticins and the heart: diverging actions elicited by signalling through prokineticin receptor-1 or -2. *BioMed Res Int*. 2009;81:3–4.

33. Cheng MY, Leslie FM, Zhou QY. Expression of prokineticins and their receptors in the adult mouse brain. *J Comp Neurol.* 2006;498:796–809.
34. Koyama Y, Kiyo-oka M, Osakada M, et al. Expression of prokineticin receptors in mouse cultured astrocytes and involvement in cell proliferation. *Brain Res.* 2006;1112:65–69.
35. Podlovni H, Ovadia O, Kisliouk T, et al. Differential expression of prokineticin receptors by endothelial cells derived from different vascular beds: a physiological basis for distinct endothelial function. *Cell Physiol Biochem.* 2006;18:315–326.
36. Li JD, Hu WP, Zhou QY. The circadian output signals from the suprachiasmatic nuclei. *Prog Brain Res.* 2012;199:119–127.
37. Moore RY. Circadian rhythms: basic neurobiology and clinical applications. *Annu Rev Med.* 1997;48:253–266.
38. Allada R, Emery P, Takahashi JS, Rosbash M. Stopping time: the genetics of fly and mouse circadian clocks. *Annu Rev Neurosci.* 2001;24:1091–1119.
39. Reppert SM, Weaver DR. Coordination of circadian timing in mammals. *Nature.* 2002;418:935–941.
40. Zhou QY, Cheng MY. Prokineticin 2 and circadian clock output. *FEBS J.* 2005;272:5703–5709.
41. Takahashi JS. Transcriptional architecture of the mammalian circadian clock. *Nat Rev Genet.* 2017;18:164–179.
42. Zhang S, Zhang J, Deng Z, et al. Circadian clock components RORalpha and Bmal1 mediate the anti-proliferative effect of MLN4924 in osteosarcoma cells. *Oncotarget.* 2016;7:66087–66099.
43. Ma YT, Luo H, Guan WJ, et al. O-GlcNAcylation of BMAL1 regulates circadian rhythms in NIH3T3 fibroblasts. *Biochem Biophys Res Commun.* 2013;431:382–387.
44. Li X, Shi L, Zhang K, et al. CirGRDB: a database for the genome-wide deciphering circadian genes and regulators. *Nucleic Acids Res.* 2018;46:D64–D70.
45. Zhang C, Truong KK, Zhou QY. Efferent projections of prokineticin 2 expressing neurons in the mouse suprachiasmatic nucleus. *PLoS One.* 2009;4:e7151.
46. Hu WP, Li JD, Zhang C, Boehmer L, Siegel JM, Zhou QY. Altered circadian and homeostatic sleep regulation in prokineticin 2-deficient mice. *Sleep.* 2007;30:247–256.
47. Prosser HM, Bradley A, Chesham JE, Ebling FJ, Hastings MH, Maywood ES. Prokineticin receptor 2 (Prokr2) is essential for the regulation of circadian behavior by the suprachiasmatic nuclei. *Proc Natl Acad Sci U S A.* 2007;104:648–653.
48. Qu Z, Zhang H, Huang M, et al. Loss of ZBTB20 impairs circadian output and leads to unimodal behavioral rhythms. *eLife.* 2016;5.
49. Chen S, Reichert S, Singh C, Oikonomou G, Rihel J, Prober DA. Light-dependent regulation of sleep and wake states by prokineticin 2 in zebrafish. *Neuron.* 2017;95:153–168. e6.
50. Zhou QY, Burton KJ, Neal ML, et al. Differential arousal regulation by prokineticin 2 signaling in the nocturnal mouse and the diurnal monkey. *Mol Brain.* 2016;9:78.
51. Negri L, Lattanzi R, Giannini E, De Felice M, Colucci A, Melchiorri P. Bv8, the amphibian homologue of the mammalian prokineticins, modulates ingestive behaviour in rats. *Br J Pharmacol.* 2004;142:181–191.
52. Cottrell GT, Zhou QY, Ferguson AV. Prokineticin 2 modulates the excitability of subfornical organ neurons. *J Neurosci.* 2004;24:2375–2379.
53. Fry M, Cottrell GT, Ferguson AV. Prokineticin 2 influences subfornical organ neurons through regulation of MAP kinase and the modulation of sodium channels. *Am J Physiol Regul Integr Comp Physiol.* 2008;295:R848–R856.
54. Gardiner JV, Bataveljic A, Patel NA, et al. Prokineticin 2 is a hypothalamic neuropeptide that potently inhibits food intake. *Diabetes.* 2010;59:397–406.

55. Chaly AL, Srisai D, Gardner EE, Sebag JA. The melanocortin receptor accessory protein 2 promotes food intake through inhibition of the prokineticin receptor-1. *eLife*. 2016;5:e12397.
56. Beale K, Gardiner JV, Bewick GA, et al. Peripheral administration of prokineticin 2 potently reduces food intake and body weight in mice via the brainstem. *Br J Pharmacol*. 2013;168:403–410.
57. Li JD, Hu WP, Zhou QY. Disruption of the circadian output molecule prokineticin 2 results in anxiolytic and antidepressant-like effects in mice. *Neuropsychopharmacology*. 2009;34:367–373.
58. Zhou W, Li JD, Hu WP, Cheng MY, Zhou QY. Prokineticin 2 is involved in the thermoregulation and energy expenditure. *Regul Pept*. 2012;179:84–90.
59. Jethwa PH, I'Anson H, Warner A, et al. Loss of prokineticin receptor 2 signaling predisposes mice to torpor. *Am J Physiol Regul Integr Comp Physiol*. 2008;294: R1968–R1979.
60. Kishi T, Kitajima T, Tsunoka T, et al. Possible association of prokineticin 2 receptor gene (PROKR2) with mood disorders in the Japanese population. *Neuromol Med*. 2009;11:114–122.
61. Kishi T, Kitajima T, Tsunoka T, et al. PROKR2 is associated with methamphetamine dependence in the Japanese population. *Prog Neuropsychopharmacol Biol Psychiatry*. 2010;34:1033–1036.
62. Negri L, Lattanzi R, Giannini E, et al. Nociceptive sensitization by the secretory protein Bv8. *Br J Pharmacol*. 2002;137:1147–1154.
63. Hu WP, Zhang C, Li JD, et al. Impaired pain sensation in mice lacking prokineticin 2. *Mol Pain*. 2006;2:35.
64. Negri L, Lattanzi R, Giannini E, et al. Impaired nociception and inflammatory pain sensation in mice lacking the prokineticin receptor PKR1: focus on interaction between PKR1 and the capsaicin receptor TRPV1 in pain behavior. *J Neurosci*. 2006;26:6716–6727.
65. Vellani V, Colucci M, Lattanzi R, et al. Sensitization of transient receptor potential vanilloid 1 by the prokineticin receptor agonist Bv8. *J Neurosci*. 2006;26:5109–5116.
66. Julius D. TRP channels and pain. *Annu Rev Cell Dev Biol*. 2013;29:355–384.
67. Xiong YC, Li XM, Wang XJ, et al. Prokineticin 2 suppresses GABA-activated current in rat primary sensory neurons. *Neuropharmacology*. 2010;59:589–594.
68. Qiu CY, Liu YQ, Qiu F, Wu J, Zhou QY, Hu WP. Prokineticin 2 potentiates acid-sensing ion channel activity in rat dorsal root ganglion neurons. *J Neuroinflammation*. 2012;9:108.
69. Ren C, Qiu CY, Gan X, et al. Prokineticin 2 facilitates mechanical allodynia induced by alpha,beta-methylene ATP in rats. *Eur J Pharmacol*. 2015;767:24–29.
70. de Novellis V, Negri L, Lattanzi R, et al. The prokineticin receptor agonist Bv8 increases GABA release in the periaqueductal grey and modifies RVM cell activities and thermoceptive reflexes in the rat. *Eur J Neurosci*. 2007;26:3068–3078.
71. Giannini E, Lattanzi R, Nicotra A, et al. The chemokine Bv8/prokineticin 2 is up-regulated in inflammatory granulocytes and modulates inflammatory pain. *Proc Natl Acad Sci U S A*. 2009;106:14646–14651.
72. Negri L, Lattanzi R. Bv8/PK2 and prokineticin receptors: a druggable pronociceptive system. *Curr Opin Pharmacol*. 2012;12:62–66.
73. Kurosaka D, Noda K, Yoshida K, et al. Elevation of Bombina variegata peptide 8 in mice with collagen-induced arthritis. *BMC Musculoskelet Disord*. 2009;10:45.
74. Ito H, Noda K, Yoshida K, et al. Prokineticin 2 antagonist, PKRA7 suppresses arthritis in mice with collagen-induced arthritis. *BMC Musculoskelet Disord*. 2016;17:387.
75. Guida F, Luongo L, Marmo F, et al. Palmitoylethanolamide reduces pain-related behaviors and restores glutamatergic synapses homeostasis in the medial prefrontal cortex of neuropathic mice. *Mol Brain*. 2015;8:47.

76. Maftei D, Marconi V, Florenzano F, et al. Controlling the activation of the Bv8/prokineticin system reduces neuroinflammation and abolishes thermal and tactile hyperalgesia in neuropathic animals. *Br J Pharmacol*. 2014;171:4850–4865.
77. Castelli M, Amodeo G, Negri L, et al. Antagonism of the prokineticin system prevents and reverses allodynia and inflammation in a mouse model of diabetes. *PLoS One*. 2016;11:e0146259.
78. Hang LH, Luo H, Li SN, et al. Involvement of spinal Bv8/prokineticin 2 in a rat model of cancer-induced bone pain. *Basic Clin Pharmacol Toxicol*. 2015;117:180–185.
79. Altman J. Autoradiographic and histological studies of postnatal neurogenesis. IV. Cell proliferation and migration in the anterior forebrain, with special reference to persisting neurogenesis in the olfactory bulb. *J Comp Neurol*. 1969;137:433–457.
80. Gage FH. Mammalian neural stem cells. *Science*. 2000;287:1433–1438.
81. Hayashi Y, Jinnou H, Sawamoto K, Hitoshi S. Adult neurogenesis and its role in brain injury and psychiatric diseases. *J Neurochem*. 2018. https://doi.org/10.1111/jnc.14557.
82. Ng KL, Li JD, Cheng MY, Leslie FM, Lee AG, Zhou QY. Dependence of olfactory bulb neurogenesis on prokineticin 2 signaling. *Science*. 2005;308:1923–1927.
83. Matsumoto S, Yamazaki C, Masumoto KH, et al. Abnormal development of the olfactory bulb and reproductive system in mice lacking prokineticin receptor PKR2. *Proc Natl Acad Sci U S A*. 2006;103:4140–4145.
84. Ragancokova D, Rocca E, Oonk AM, et al. TSHZ1-dependent gene regulation is essential for olfactory bulb development and olfaction. *Sci Rep*. 2014;124:1214–1227.
85. Nebigil CG. Prokineticin is a new linker between obesity and cardiovascular diseases. *Front Cardiovas Med*. 2017;4:20.
86. Urayama K, Guilini C, Messaddeq N, et al. The prokineticin receptor-1 (GPR73) promotes cardiomyocyte survival and angiogenesis. *FASEB J*. 2007;21:2980–2993.
87. Arora H, Boulberdaa M, Qureshi R, et al. Prokineticin receptor-1 signaling promotes epicardial to mesenchymal transition during heart development. *Sci Rep*. 2016;6:25541.
88. Gasser A, Brogi S, Urayama K, et al. Discovery and cardioprotective effects of the first non-peptide agonists of the G protein-coupled prokineticin receptor-1. *PLoS One*. 2015;10:e0121027.
89. LeCouter J, Ferrara N. EG-VEGF and Bv8. A novel family of tissue-selective mediators of angiogenesis, endothelial phenotype, and function. *Trends Cardiovasc Med*. 2003;13:276–282.
90. Monnier J, Samson M. Prokineticins in angiogenesis and cancer. *Cancer Lett*. 2010;296:144–149.
91. Samson M, Peale Jr FV, Frantz G, Rioux-Leclercq N, Rajpert-De Meyts E, Ferrara N. Human endocrine gland-derived vascular endothelial growth factor: expression early in development and in Leydig cell tumors suggests roles in normal and pathological testis angiogenesis. *J Clin Endocrinol Metab*. 2004;89:4078–4088.
92. Brouillet S, Hoffmann P, Feige JJ, Alfaidy N. EG-VEGF: a key endocrine factor in placental development. *Trends Endocrinol Metab*. 2012;23:501–508.
93. Fraser HM, Bell J, Wilson H, et al. Localization and quantification of cyclic changes in the expression of endocrine gland vascular endothelial growth factor in the human corpus luteum. *J Clin Endocrinol Metab*. 2005;90:427–434.
94. Hoffmann P, Feige JJ, Alfaidy N. Expression and oxygen regulation of endocrine gland-derived vascular endothelial growth factor/prokineticin-1 and its receptors in human placenta during early pregnancy. *Endocrinology*. 2006;147:1675–1684.
95. Gao MZ, Zhao XM, Sun ZG, Hong Y, Zhao LW, Zhang HQ. Endocrine gland-derived vascular endothelial growth factor concentrations in follicular fluid and serum may predict ovarian hyperstimulation syndrome in women undergoing controlled ovarian hyperstimulation. *Fertil Steril*. 2011;95:673–678.

96. Haouzi D, Mahmoud K, Fourar M, et al. Identification of new biomarkers of human endometrial receptivity in the natural cycle. *Hum Reprod.* 2009;24:198–205.
97. Evans J, Catalano RD, Brown P, et al. Prokineticin 1 mediates fetal-maternal dialogue regulating endometrial leukemia inhibitory factor. *FASEB J.* 2009;23:2165–2175.
98. Hoffmann P, Saoudi Y, Benharouga M, et al. Role of EG-VEGF in human placentation: physiological and pathological implications. *J Cell Mol Med.* 2009;13:2224–2235.
99. Brouillet S, Hoffmann P, Benharouga M, et al. Molecular characterization of EG-VEGF-mediated angiogenesis: differential effects on microvascular and macrovascular endothelial cells. *Mol Biol Cell.* 2010;21:2832–2843.
100. Dunand C, Hoffmann P, Sapin V, et al. Endocrine gland-derived endothelial growth factor (EG-VEGF) is a potential novel regulator of human parturition. *Biol Reprod.* 2014;91:73.
101. Catalano RD, Lannagan TR, Gorowiec M, Denison FC, Norman JE, Jabbour HN. Prokineticins: novel mediators of inflammatory and contractile pathways at parturition? *Mol Hum Reprod.* 2010;16:311–319.
102. Shaw JL, Denison FC, Evans J, et al. Evidence of prokineticin dysregulation in fallopian tube from women with ectopic pregnancy. *Fertil Steril.* 2010;94:1601–1608 [e1].
103. Salker M, Teklenburg G, Molokhia M, et al. Natural selection of human embryos: impaired decidualization of endometrium disables embryo-maternal interactions and causes recurrent pregnancy loss. *PLoS One.* 2010;5:e10287.
104. Su MT, Lin SH, Chen YC, Kuo PL. Gene–gene interactions and risk of recurrent miscarriages in carriers of endocrine gland-derived vascular endothelial growth factor and prokineticin receptor polymorphisms. *Fertil Steril.* 2014;102:1071–1077 [e3].
105. Su MT, Lin SH, Chen YC, Wu LW, Kuo PL. Prokineticin receptor variants (PKR1-I379V and PKR2-V331M) are protective genotypes in human early pregnancy. *Reproduction.* 2013;146:63–73.
106. Su MT, Lin SH, Lee IW, et al. Polymorphisms of endocrine gland-derived vascular endothelial growth factor gene and its receptor genes are associated with recurrent pregnancy loss. *Hum Reprod.* 2010;25:2923–2930.
107. Brouillet S, Murthi P, Hoffmann P, et al. EG-VEGF controls placental growth and survival in normal and pathological pregnancies: case of fetal growth restriction (FGR). *Cell Mol Life Sci.* 2013;70:511–525.
108. Shinkaruk S, Bayle M, Lain G, Deleris G. Vascular endothelial cell growth factor (VEGF), an emerging target for cancer chemotherapy. *Curr Med Chem Anticancer Agents.* 2003;3:95–117.
109. Corlan AS, Cimpean AM. Endocrine gland-derived vascular endothelial growth factor/prokineticin-1 in cancer development and tumor angiogenesis. *Intl J Endocrinol.* 2017;2017:3232905.
110. Diaz-Montero CM, Salem ML, Nishimura MI, Garrett-Mayer E, Cole DJ, Montero AJ. Increased circulating myeloid-derived suppressor cells correlate with clinical cancer stage, metastatic tumor burden, and doxorubicin-cyclophosphamide chemotherapy. *Cancer Immunol Immunother.* 2009;58:49–59.
111. Marigo I, Dolcetti L, Serafini P, Zanovello P, Bronte V. Tumor-induced tolerance and immune suppression by myeloid derived suppressor cells. *Immunol Rev.* 2008;222:162–179.
112. Serafini P, Borrello I, Bronte V. Myeloid suppressor cells in cancer: recruitment, phenotype, properties, and mechanisms of immune suppression. *Semin Cancer Biol.* 2006;16:53–65.
113. Yang L, DeBusk LM, Fukuda K, et al. Expansion of myeloid immune suppressor Gr+CD11b+ cells in tumor-bearing host directly promotes tumor angiogenesis. *Cancer Cell.* 2004;6:409–421.

114. Chung AS, Ferrara N. Developmental and pathological angiogenesis. *Annu Rev Cell Dev Biol.* 2011;27:563–584.
115. Shojaei F, Zhong C, Wu X, Yu L, Ferrara N. Role of myeloid cells in tumor angiogenesis and growth. *Trends Cell Biol.* 2008;18:372–378.
116. Shojaei F, Wu X, Malik AK, et al. Tumor refractoriness to anti-VEGF treatment is mediated by CD11b+Gr1+ myeloid cells. *Nat Biotechnol.* 2007;25:911–920.
117. Shojaei F, Wu X, Qu X, et al. G-CSF-initiated myeloid cell mobilization and angiogenesis mediate tumor refractoriness to anti-VEGF therapy in mouse models. *Proc Natl Acad Sci U S A.* 2009;106:6742–6747.
118. Shojaei F, Wu X, Zhong C, et al. Bv8 regulates myeloid-cell-dependent tumour angiogenesis. *Nature.* 2007;450:825–831.
119. Cheng MY, Lee AG, Culbertson C, et al. Prokineticin 2 is an endangering mediator of cerebral ischemic injury. *Proc Natl Acad Sci U S A.* 2012;109:5475–5480.
120. Landucci E, Lattanzi R, Gerace E, et al. Prokineticins are neuroprotective in models of cerebral ischemia and ischemic tolerance in vitro. *Neuropharmacology.* 2016;108: 39–48.
121. Gordon R, Neal ML, Luo J, Langley MR, Harischandra DS. Prokineticin-2 upregulation during neuronal injury mediates a compensatory protective response against dopaminergic neuronal degeneration. *Nat Commun.* 2016;7, 12932.
122. Caioli S, Severini C, Ciotti T, et al. Prokineticin system modulation as a new target to counteract the amyloid beta toxicity induced by glutamatergic alterations in an in vitro model of Alzheimer's disease. *Neuropharmacology.* 2017;116:82–97.
123. Severini C, Lattanzi R, Maftei D, et al. Bv8/prokineticin 2 is involved in Abeta-induced neurotoxicity. *Sci Rep.* 2015;5:15301.
124. Seminara SB, Hayes FJ, Crowley Jr WF. Gonadotropin-releasing hormone deficiency in the human (idiopathic hypogonadotropic hypogonadism and Kallmann's syndrome): pathophysiological and genetic considerations. *Endocr Rev.* 1998;19:521–539.
125. Boehm U, Bouloux PM, Dattani MT, et al. Expert consensus document: European consensus statement on congenital hypogonadotropic hypogonadism—pathogenesis, diagnosis and treatment. *Nat Rev Endocrinol.* 2015;11:547–564.
126. Mitchell AL, Dwyer A, Pitteloud N, Quinton R. Genetic basis and variable phenotypic expression of Kallmann syndrome: towards a unifying theory. *Trends Endocrinol Metab.* 2011;22:249–258.
127. Reitano JF, Caminos-Torres R, Snyder PJ. Serum LH and FSH responses to the repetitive administration of gonadotropin-releasing hormone in patients with idiopathic hypogonadotropic hypogonadism. *J Clin Endocrinol Metab.* 1975;41: 1035–1042.
128. Bianco SD, Kaiser UB. The genetic and molecular basis of idiopathic hypogonadotropic hypogonadism. *Nat Rev Endocrinol.* 2009;5:569–576.
129. Waldstreicher J, Seminara SB, Jameson JL, et al. The genetic and clinical heterogeneity of gonadotropin-releasing hormone deficiency in the human. *J Clin Endocrinol Metab.* 1996;81:4388–4395.
130. Quinton R, Duke VM, Robertson A, et al. Idiopathic gonadotrophin deficiency: genetic questions addressed through phenotypic characterization. *Clin Endocrinol (Oxf).* 2001;55:163–174.
131. Gutierrez-Amavizca BE, Figuera LE, Orozco-Castellanos R. Current genetic issues and phenotypic variants in Kallmann syndrome. *Rev Med Inst Mex Seguro Soc.* 2012;50:157–161.
132. Brioude F, Bouligand J, Trabado S, et al. Non-syndromic congenital hypogonadotropic hypogonadism: clinical presentation and genotype–phenotype relationships. *Eur J Endocrinol.* 2010;162:835–851.

133. Conn PM, Crowley Jr WF. Gonadotropin-releasing hormone and its analogs. *Annu Rev Med*. 1994;45:391–405.
134. Wetsel WC, Srinivasan S. Pro-GnRH processing. *Prog Brain Res*. 2002;141:221–241.
135. Cano Sokoloff N, Misra M, Ackerman KE. Exercise, training, and the hypothalamic–pituitary–gonadal axis in men and women. *Front Horm Res*. 2016;47:27–43.
136. Forni PE, Wray S. GnRH, anosmia and hypogonadotropic hypogonadism—where are we? *Front Neuroendocrinol*. 2015;36:165–177.
137. Daikoku S. Olfactory origin of luteinizing hormone-releasing hormone (LHRH) neurons. *Nihon Ika Daigaku Zasshi*. 1999;66:94–106.
138. Daikoku S. The olfactory origin of luteinizing hormone-releasing hormone (LHRH) neurons. A new era in reproduction physiology. *Arch Histol Cytol*. 1999;62:107–117.
139. Murakami S, Kikuyama S, Arai Y. The origin of the luteinizing hormone-releasing hormone (LHRH) neurons in newts (*Cynops pyrrhogaster*): the effect of olfactory placode ablation. *Cell Tissue Res*. 1992;269:21–27.
140. Schwanzel-Fukuda M. Origin and migration of luteinizing hormone-releasing hormone neurons in mammals. *Microsc Res Tech*. 1999;44:2–10.
141. Schwanzel-Fukuda M, Pfaff DW. Origin of luteinizing hormone-releasing hormone neurons. *Nature*. 1989;338:161–164.
142. Pitteloud N, Zhang C, Pignatelli D, et al. Loss-of-function mutation in the prokineticin 2 gene causes Kallmann syndrome and normosmic idiopathic hypogonadotropic hypogonadism. *Proc Natl Acad Sci U S A*. 2007;104:17447–17452.
143. Dode C, Teixeira L, Levilliers J, et al. Kallmann syndrome: mutations in the genes encoding prokineticin-2 and prokineticin receptor-2. *PLoS Genet*. 2006;2:e175.
144. Cole LW, Sidis Y, Zhang C, et al. Mutations in prokineticin 2 and prokineticin receptor 2 genes in human gonadotrophin-releasing hormone deficiency: molecular genetics and clinical spectrum. *J Clin Endocrinol Metab*. 2008;93:3551–3559.
145. Sbai O, Monnier C, Dode C, Pin JP, Hardelin JP, Rondard P. Biased signaling through G-protein-coupled PROKR2 receptors harboring missense mutations. *FASEB J*. 2014;28:3734–3744.
146. Libri DV, Kleinau G, Vezzoli V, et al. Germline prokineticin receptor 2 (PROKR2) variants associated with central hypogonadism cause differential modulation of distinct intracellular pathways. *J Clin Endocrinol Metab*. 2014;99:E458–E463.
147. Dode C, Rondard P. PROK2/PROKR2 signaling and Kallmann syndrome. *Front Endocrinol*. 2013;4:19.
148. Monnier C, Dode C, Fabre L, et al. PROKR2 missense mutations associated with Kallmann syndrome impair receptor signalling activity. *Hum Mol Genet*. 2009;18:75–81.
149. Raivio T, Avbelj M, McCabe MJ, et al. Genetic overlap in Kallmann syndrome, combined pituitary hormone deficiency, and septo-optic dysplasia. *J Clin Endocrinol Metab*. 2012;97:E694–E699.
150. Abreu AP, Kaiser UB, Latronico AC. The role of prokineticins in the pathogenesis of hypogonadotropic hypogonadism. *Neuroendocrinology*. 2010;91:283–290.
151. Martin C, Balasubramanian R, Dwyer AA, et al. The role of the prokineticin 2 pathway in human reproduction: evidence from the study of human and murine gene mutations. *Endocr Rev*. 2011;32:225–246.
152. Quaynor SD, Kim HG, Cappello EM, et al. The prevalence of digenic mutations in patients with normosmic hypogonadotropic hypogonadism and Kallmann syndrome. *Fertil Steril*. 2011;96:1424–1430. e6.
153. Shaw ND, Seminara SB, Welt CK, et al. Expanding the phenotype and genotype of female GnRH deficiency. *J Clin Endocrinol Metab*. 2011;96:E566–E576.
154. Sarfati J, Guiochon-Mantel A, Rondard P, et al. A comparative phenotypic study of Kallmann syndrome patients carrying monoallelic and biallelic mutations in the prokineticin 2 or prokineticin receptor 2 genes. *J Clin Endocrinol Metab*. 2010;95: 659–669.

155. Costa-Barbosa FA, Balasubramanian R, Keefe KW, et al. Prioritizing genetic testing in patients with Kallmann syndrome using clinical phenotypes. *J Clin Endocrinol Metab.* 2013;98:E943–E953.
156. Huijbregts L, Roze C, Bonafe G, et al. DNA polymorphisms of the KiSS1 3′ untranslated region interfere with the folding of a G-rich sequence into G-quadruplex. *Mol Cell Endocrinol.* 2012;351:239–248.
157. Tommiska J, Toppari J, Vaaralahti K, et al. PROKR2 mutations in autosomal recessive Kallmann syndrome. *Fertil Steril.* 2013;99:815–818.
158. Bergman JE, de Ronde W, Jongmans MC, et al. The results of CHD7 analysis in clinically well-characterized patients with Kallmann syndrome. *J Clin Endocrinol Metab.* 2012;97:E858–E862.
159. Chan YM, de Guillebon A, Lang-Muritano M, et al. GNRH1 mutations in patients with idiopathic hypogonadotropic hypogonadism. *Proc Natl Acad Sci U S A.* 2009;106:11703–11708.
160. Abreu AP, Trarbach EB, de Castro M, et al. Loss-of-function mutations in the genes encoding prokineticin-2 or prokineticin receptor-2 cause autosomal recessive Kallmann syndrome. *J Clin Endocrinol Metab.* 2008;93:4113–4118.
161. Cox KH, Oliveira LMB, Plummer L, et al. Modeling mutant/wild-type interactions to ascertain pathogenicity of PROKR2 missense variants in patients with isolated GnRH deficiency. *Hum Mol Genet.* 2018;27:338–350.
162. Sarfati J, Fouveaut C, Leroy C, Jeanpierre M, Hardelin JP, Dode C. Greater prevalence of PROKR2 mutations in Kallmann syndrome patients from the Maghreb than in European patients. *Eur J Endocrinol.* 2013;169:805–809.
163. Tao YX. Inactivating mutations of G protein-coupled receptors and diseases: structure–function insights and therapeutic implications. *Pharmacol Ther.* 2006;111: 949–973.
164. Zhou XT, Chen DN, Xie ZQ, et al. Functional analysis of the distal region of the third intracellular loop of PROKR2. *Biochem Biophys Res Commun.* 2013;439:12–17.
165. Sykiotis GP, Plummer L, Hughes VA, et al. Oligogenic basis of isolated gonadotropin-releasing hormone deficiency. *Proc Natl Acad Sci U S A.* 2010;107:15140–15144.
166. Avbelj Stefanija M, Jeanpierre M, Sykiotis GP, et al. An ancient founder mutation in PROKR2 impairs human reproduction. *Hum Mol Genet.* 2012;21:4314–4324.
167. McCabe MJ, Gaston-Massuet C, Gregory LC, et al. Variations in PROKR2, but not PROK2, are associated with hypopituitarism and septo-optic dysplasia. *J Clin Endocrinol Metab.* 2013;98:E547–E557.
168. Reynaud R, Jayakody SA, Monnier C, et al. PROKR2 variants in multiple hypopituitarism with pituitary stalk interruption. *J Clin Endocrinol Metab.* 2012;97: E1068–E1073.
169. Loucks AB, Mortola JF, Girton L, Yen SS. Alterations in the hypothalamic–pituitary–ovarian and the hypothalamic–pituitary–adrenal axes in athletic women. *J Clin Endocrinol Metab.* 1989;68:402–411.
170. Giles DE, Berga SL. Cognitive and psychiatric correlates of functional hypothalamic amenorrhea: a controlled comparison. *Fertil Steril.* 1993;60:486–492.
171. Caronia LM, Martin C, Welt CK, et al. A genetic basis for functional hypothalamic amenorrhea. *N Engl J Med.* 2011;364:215–225.

CHAPTER NINE

Therapeutic strategies for diseases caused by loss-of-function mutations in G protein-coupled receptors

Wei Wang[a], **Dong-Yu Guo**[a,*], **Ya-Xiong Tao**[b,*]

[a]Department of Clinical Laboratory, Xiamen Huli Guoyu Clinic, Co., Ltd., Xiamen, China
[b]Department of Anatomy, Physiology, and Pharmacology, College of Veterinary Medicine, Auburn University, Auburn, AL, United States
*Corresponding authors: e-mail address: xiamenhaijin@163.com; taoyaxi@auburn.edu

Contents

Abstract

As one of the largest families of cell membrane proteins, G protein-coupled receptors (GPCRs) are involved in regulating almost all physiological processes by transducing extracellular signals into the cytoplasm. Since the first discovery of naturally occurring mutations in *Rhodopsin* gene in 1990, hundreds of loss-of-function mutations in multiple GPCRs have been identified to be pathogenic for more than 30 diverse human

Progress in Molecular Biology and Translational Science, Volume 161
ISSN 1877-1173
https://doi.org/10.1016/bs.pmbts.2018.09.002

diseases, making these defective receptors important drug targets for personalized medicine. In this review, we aim to elucidate the etiologies of five common inherited diseases caused by six of the most extensively studied GPCRs. The molecular basis and classification of inactivating mutations in GPCRs are also reviewed. The available therapeutic approaches directed against different classes of mutants, especially pharmacological chaperones targeting intracellularly retained mutants, reported during the past two decades, are systematically summarized.

Abbreviations

AAV	adeno-associated virus
adRP	autosomal dominant retinitis pigmentosa
AVPR2	arginine vasopressin receptor 2 (gene name for V2R)
DMSO	dimethyl sulfoxide
ER	endoplasmic reticulum
ERG	electroretinogram
FSH	follicle-stimulating hormone
FSHR	follicle-stimulating hormone receptor
GnRHR	gonadotropin-releasing hormone receptor
GPCR	G protein-coupled receptor
hCG	human chorionic gonadotropin
LH	luteinizing hormone
LHCGR	luteinizing hormone/chorionic gonadotropin receptor
MC4R	melanocortin-4 receptor
NDI	nephrogenic diabetes insipidus
4-PBA	4-phenylbutyrate
PC	pharmacological chaperone
PTC	premature termination codon
RHO	rhodopsin
TM	transmembrane
TMAO	trimethyl-amine-*N*-oxide
TUDCA	tauroursodeoxycholic acid
V2R	V2 vasopressin receptor
WT	wild type

1. Introduction

As one of the largest and most functionally diverse family of membrane proteins, G protein-coupled receptors (GPCRs) act as signal transducers to transduce an array of both external stimuli (such as photons, odorants, and pheromones) and internal stimuli (such as ions, lipids, and hormones) into the cell. All GPCRs share the same topology consisting of seven α-helical transmembrane (TM) domains connected by alternating extracellular and intracellular loops and are flanked by extracellular N-termini and intracellular C-terminal tails. Since GPCRs regulate almost all physiological functions, they have been popular therapeutic targets, with an estimate of ~27% of current drugs in the global market targeting GPCRs, accounting for ~\$890 billion in sales from 2011 to 2015.[1]

Since the first discovery of naturally occurring mutations in *Rhodopsin* (*RHO*) in 1990,[2] hundreds of loss-of-function (inactivating) mutations in multiple GPCRs have been identified to be pathogenic for more than 30 diverse human diseases.[3,4] Some of the most extensively studied GPCRs in this aspect are rhodopsin, V2 vasopressin receptor (V2R), melanocortin-4 receptor (MC4R), gonadotropin-releasing hormone (GnRH) receptor (GnRHR), luteinizing hormone (LH)/chorionic gonadotropin (CG) receptor (LHCGR), and follicle-stimulating hormone (FSH) receptor (FSHR). In this chapter, we seek to illustrate the molecular basis, pathophysiological mechanisms, and therapeutic approaches for inherited diseases resulted from inactivating GPCR mutations.

2. Selected diseases caused by loss-of-function mutations in G protein-coupled receptors

2.1 *Rhodopsin* mutations and retinitis pigmentosa

Rhodopsin, the dim-light photoreceptor expressed in the rod cells of the retina, is a crucial player in GPCR structural biology when its high-resolution crystal structure was first reported in 2000.[5] Absorption of light by 11-*cis*-retinal chromophore results in rapid photochemical isomerization and receptor activation, through which rhodopsin is converted to its active form, metarhodopsin II, to induce the downstream phototransduction signaling pathway.[6] The 11-*cis*-retinal also acts as an inverse agonist to suppress constitutive activity of the receptor. The first GPCR mutation pathogenic for human disease was reported by Dryja et al., who demonstrated that P23H *RHO* causes human autosomal dominant retinitis pigmentosa (adRP),[2] a degenerative retinal disease usually begins with night blindness and progresses to peripheral vision loss and even total blindness. Since then, more than 150 distinct *RHO* mutations have been shown to be associated with adRP.[4] Functional characterization of these mutants revealed that the majority of the mutant receptors are misfolded and trapped intracellularly, prone to be degraded by the ubiquitin–proteasome pathway.[7,8]

2.2 *AVPR2* mutations and nephrogenic diabetes insipidus

Vasopressin, a peptide hormone synthesized in the hypothalamus and released in response of increased blood osmolarity or decreased cardiac volume, plays an important role in renal water reabsorption to maintain water homeostasis. This antidiuretic effect is achieved by activation of its receptor, V2R, via the Gs–adenylyl cyclase–cAMP-dependent signaling, leading to translocation of water channel aquaporin-2 to the apical membrane. Mutations in the *AVPR2* (*arginine vasopressin receptor 2*) gene (encoding V2R) are the major cause for X-linked nephrogenic diabetes insipidus (NDI), a rare inherited disease characterized by failure to respond to vasopressin and to concentrate urine.[9] The clinical symptoms include polyuria, polydipsia, and hyposthenuria. To date, more than 200 *AVPR2* mutations have been reported in X-linked NDI patients, with intracellular retention being the most common defect.[10]

2.3 *MC4R* mutations and obesity

It is well established that the MC4R, a downstream mediator of leptin action in the hypothalamus, plays an important role in the regulation of energy homeostasis in both rodents and humans. As a rhodopsin-like GPCR, the MC4R primarily responds to the activation of a potent endogenous peptide agonist, α-melanocyte-stimulating hormone (α-MSH), resulting in increased satiety, energy expenditure, and weight loss, and to the inhibition of a natural antagonist/inverse agonist, agouti-related peptide, leading to increased food intake, energy conservation, and weight gain. Since the first 2 reports of frameshift *MC4R* mutations that correlate with early-onset morbid obesity,[11,12] approximately 170 *MC4R* mutations have been identified in obese patients,[13,14] representing the most common monogenic form of obesity with a prevalence of up to 6% in some ethnic populations.[15,16]

2.4 *GNRHR* mutations and hypogonadotropic hypogonadism

GnRH and its receptor, GnRHR, play critical roles in the hypothalamus–pituitary–gonadal axis. Receptor activation by GnRH stimulates the secretion of gonadotropins, resulting in development and maturation of gonads in fetal life and after birth. Inactivating mutations in the *GNRHR* gene resulting in hypogonadotropic hypogonadism, a reproductive disease characterized by low levels of gonadotropins and delayed or absent sexual development, was originally reported by de Roux et al.[17] Since then, more than 31 *GNRHR* mutations have been described in the literature (reviewed in Refs. 18,19).

Hypogonadotropic hypogonadism is transmitted as a recessive trait in all published cases of inactivating *GNRHR* mutations to date.

2.5 *LHCGR* and *FSHR* mutations and hypergonadotropic hypogonadism

The LHCGR and FSHR, collectively termed gonadotropin receptors, are pivotal mediators in reproductive physiology. As two structurally relevant rhodopsin-like GPCRs, the endogenous ligands for LHCGR (LH and CG) and FSHR (FSH) are composed of two subunits, an identical α-subunit and a hormone-specific β-subunit.[20] LH and FSH are synthesized in gonadotroph cells of the anterior pituitary in response to hypothalamic release of GnRH, while human chorionic gonadotropin (hCG) is secreted from the placenta of pregnant women. In males, activation of LHCGR expressed in Leydig cells by LH during puberty results in the production and secretion of androgens, which are essential for male sexual development, whereas activation of FSHR expressed on Sertoli cells stimulates spermatogenesis. In females, activation of LHCGR expressed on ovarian theca cells stimulates androgen synthesis, induces ovulation of Graafian follicles, and stimulates progesterone synthesis by luteal cells, while activation of FSHR expressed on granulosa cells induces the aromatization of androgens to estrogen, follicular growth, and inhibin secretion.[21,22]

Therefore, loss-of-function mutations in these receptors have been implicated in several reproductive disorders, including Leydig cell hypoplasia and hypergonadotropic hypogonadism in males, as well as polycystic ovaries, amenorrhea, and infertility in females.[23,24] To date, approximately 34 and 18 naturally occurring inactivating mutations of the *LHCGR* and *FSHR* genes have been described in patients suffering from varying degrees of reproductive dysfunction, respectively.[24–26]

3. Molecular basis of loss-of-function mutations in G protein-coupled receptors

3.1 Life cycle of G protein-coupled receptors

It is well established that GPCRs begin their life cycle from transcription into mRNA in the nucleus, followed by translation into polypeptides in the ribosome located at rough endoplasmic reticulum (ER) and inserted into the ER, where the nascent proteins are scrutinized by the stringent cell quality control system. Only correctly folded receptors are able to route onto the cell membrane after further processing in other cell compartments, especially the Golgi apparatus. At the cell surface, GPCRs bind to ligands and undergo conformational changes, resulting in downstream signal transduction by coupling and catalyzation of G proteins. Prolonged activation of GPCRs usually leads to receptor desensitization and internalization. The internalized GPCRs will either recycle back to cell membrane or degrade in the lysosome (reviewed in Ref. 27).

3.2 Classification of loss-of-function mutations in G protein-coupled receptors

Depending on the functional defect occurred in the process described earlier, mutants can be cataloged into five classes according to a classification scheme we originally proposed for the MC4R and then applied to other GPCRs.[4,28–30] Class I mutants are defective in receptor biosynthesis and/or have accelerated protein degradation, resulting in decreased total receptor protein; Class II mutants are synthesized but trapped intracellularly, primarily by the ER quality control system, although some mutants may be trapped in the Golgi apparatus; Class III mutants are defective in ligand binding; Class IV mutants are defective in signaling; and Class V mutants manifest with normal function compared with the corresponding wild-type (WT) receptor (Fig. 1).

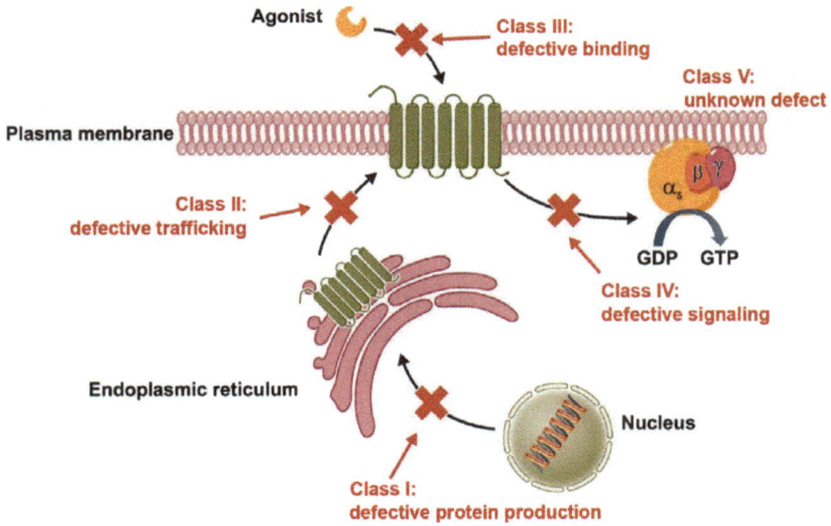

Fig. 1 Schematic classification of naturally occurring loss-of-function mutations in GPCRs. *From Tao YX, Conn PM. Chaperoning G protein-coupled receptors: from cell biology to therapeutics.* Endocr Rev. *2014;35:602–647. Copyright 2014, Oxford University Press.*

4. Therapeutic strategies

Classification of loss-of-function mutations in GPCRs not only explains the underlying molecular mechanisms responsible for receptor-related diseases but also provides insights toward personalized therapies for patients harboring certain inactivating mutations. To date, several therapeutic strategies, including aminoglycosides for Class I mutations, chemical and pharmacological chaperones for Class II mutations, novel ligands for Class III and IV mutations, as well as gene therapies, are being developed.

4.1 Aminoglycosides for Class I mutations

The formation of premature termination codons (PTCs) and subsequent truncated and nonfunctional receptors, generated from nonsense or frame-shift mutations comprises the majority of Class I mutants. Nonsense mutations account for about 11% of all described gene lesions leading to inherited human diseases, including cystic fibrosis, Duchenne muscular dystrophy, ataxia-telangiectasia, hemophilia, and some types of cancers.[31–36] Over the past decades, pharmacological agents with the ability to promote PTC suppression or translational read-through have been proposed and tested in extensive in vitro and in vivo studies and clinical trials. Among which, aminoglycoside antibiotics are the mainstay of such attempts with the most prominence.

4.1.1 V2R

The potential of utilizing aminoglycoside antibiotics for the treatment of diseases caused by nonsense GPCR mutations was first investigated by Schoneberg and colleagues.[37] In COS-7 cells transiently transfected with three truncated V2Rs, W200X, E242X, and R337X, geneticin restores vasopressin-induced cAMP production of truncated receptors by mediating a significant increase in the formation of full-length receptors,[37,38] indicating that the regain of function in truncated V2Rs is due to a read-through of the PTC.

To investigate the in vivo efficacy of geneticin in rescuing the functionality of Class I mutants, the same group generated a transgenic mouse model containing E242X in the *Avpr2* gene that mimics clinical symptoms of human X-linked NDI.[38] Intraperitoneal administration of geneticin results in a 50% increase on the urine osmolality in heterozygous mice but no significant effect on that of WT mice.[38] Further comparison study showed that geneticin is more potent (about 10 times) and effective (more than threefold) than gentamicin and paromomycin in rescuing the function of E242X V2R.[38] However, additional toxicity data obtained from transgenic mice showed that LD_{50} (defined as drug dose required to result in half death of tested population) values of geneticin, gentamicin, and paromomycin are 140, 47, and 90 mg/kg, respectively, indicating relatively high toxicity of geneticin.[38]

4.1.2 MC4R

In the MC4R, Biebermann and colleagues investigated the effect of geneticin on four obesity-associated nonsense MC4R mutations, W16X, Y35X, E61X, and Q307X.[39] In transfected COS-7 cells, geneticin treatment significantly increases the production and cell surface expression of the mutant receptors to various degrees (from high to low: W16X, E61X, Q307X, and Y35X), and the restored mutant receptors except E61X MC4R at cell membrane respond to α-MSH with robust cAMP production.[39]

4.1.3 Second-generation aminoglycosides

Due to decreased efficacy at subtoxic doses and high toxicity of conventional aminoglycosides, the clinical application of PTC suppression therapy is limited. With the recently solved X-ray crystal structure of geneticin-bound eukaryotic ribosome, Baasov and colleagues discovered a new pharmacophore 7′-hydroxyl group that increases eukaryotic selectivity and PTC read-through activity, and then designed lead compounds NB156 and NB157 as novel read-through inducers.[40] Using coupled transcription/translation assays in vitro, they showed the $IC_{50}s^{Euk}$ (half-maximal eukaryote inhibition concentration) of gentamicin, NB156, and NB157 are 62, 13.9, and 1.2 μM, respectively,[40] suggesting that NB157 is more effective in suppressing eukaryotic translation than NB156 and gentamicin. Since the correlation of reduced prokaryotic ribosome specificity with decreased ototoxic effects has been demonstrated in a systemic comparison of NB74, NB124, gentamicin, and geneticin,[41,42] NB156 and NB157 might also be less ototoxic than the conventional aminoglycosides.[40] With higher efficacy and lower toxicity profile compared to the conventional aminoglycosides, NB156 and NB157 might be utilized as promising therapeutics for treating diseases resulting from nonsense mutations.

4.2 Chemical chaperones for Class II mutants

Chemical chaperones are small and low molecular weight molecules that nonselectively promote protein folding. Based on the chemical structures and ways of action, they are divided into two groups, osmolytes and hydrophobic chaperones. Osmolytes, such as glycerol, tauroursodeoxycholic acid (TUDCA), dimethyl sulfoxide (DMSO), and trimethyl-amine-*N*-oxide (TMAO), may prevent the aggregation of misfolded or partially folded intermediates by sequestering hydrophobic regions.[43] Hydrophobic chemical chaperones such as 4-phenylbutyrate (4-PBA) can upregulate endogenous molecular chaperones by inducing a stress response and/or promote correct folding of misfolded proteins by binding to the exposed hydrophobic segments.[44]

The therapeutic potential of chemical chaperones in treating genetic diseases caused by misfolded proteins, including cystic fibrosis transmembrane conductance regulator,[45] α1-antitrypsin,[46] aquaporin-2,[47] preproparathyroid hormone,[48] and bone morphogenetic protein type II receptor,[49] and some GPCRs has been extensively explored.

4.2.1 Rhodopsin

In an attempt to alleviate toxic effects of P23H rhodopsin by different drugs, Mendes and Cheetham demonstrated that chemical chaperones, including 4-PBA, DMSO, TMAO, and trehalose, inhibit rhodopsin aggregation and inclusion formation in human SK-N-SH neuroblastoma cells.[50] 4-PBA was also shown to inhibit ER stress and apoptosis induced by misfolded T17M rhodopsin in transfected retinal pigment epithelia-19 cells.[51] However, the in vivo efficacy of 4-PBA in improving photoreceptor function or ameliorating degeneration, as assessed by electroretinogram (ERG) and retinal histology of P23H transgenic rats, was not observed.[52] On the contrary, another chemical chaperone, curcumin, was shown to improve rhodopsin function and morphology in transgenic rats, likely through suppressing P23H aggregation and ER stress and aiding rhodopsin trafficking to the rod outer segment.[53]

The beneficial effects of TUDCA have also been investigated in several retinal degeneration models. TUDCA treatment preserves cone and rod structure and function in both slow and fast degenerative P23H rat models of rhodopsin retinitis pigmentosa, as evidenced by more photoreceptors, higher amplitude of ERG a-waves (generated by photoreceptors) and b-waves (generated by secondary neurons), and smaller TUNEL-positive cells in TUDCA-treated rats than in control rats.[52,54] Unfortunately, the use of TUDCA in humans might be limited by the potential side effects associated with high efficacious dose (500 mg/kg) and the untolerated taste. To avoid these undesired effects, a safe and long-term intraocular delivery of

TUDCA by using biodegradable microspheres containing traces of TUDCA was shown with positive results in P23H transgenic rats,[55] paving the road to investigate the effect of this drug in clinical trials.

4.2.2 V2R

In the V2R, Robben et al. showed that, of nine misfolded V2R mutants tested, only V206D has increased receptor maturation and rescued plasma membrane expression upon treatment with glycerol or DMSO but not TMAO or 4-PBA,[56] indicating that the rescue is mutant- and compound-specific. Further western blot studies demonstrated that glycerol or DMSO treatment does not affect the expression of molecular chaperones such as calnexin, HSP70, or HSP90,[56] suggesting that upregulation of molecular chaperones is not involved in the rescue. Similarly, TMAO or DMSO treatment significantly restores the cell surface expression of three misrouted V2R mutants, A98P, L274P, and R113W.[57] With a large portion of naturally occurring *AVPR2* mutations being misfolded in the epithelial cells of the collecting duct, chemical chaperone therapy with the ability to promote receptor folding and trafficking might represent a promising therapeutic strategy for the treatment of X-linked NDI. However, further studies to demonstrate the efficacy and safety of these compounds in vivo and in clinical trials are needed.

4.2.3 MC4R

In the MC4R, Baldini and colleagues demonstrated that five obesity-linked MC4R variants, when expressed in immortalized neuronal cells, have significantly impaired cell surface expression, higher levels of molecular chaperones, and greater extent of ubiquitination than WT receptor.[58,59] Upon 48-h treatment of 4-PBA, the cell surface expression of WT, P78L, I316S, I317T, and P272L MC4R increases approximately by 70%, 24%, 48%, 41%, and 40%, respectively.[58,59] The rescued I316S MC4R is functional at cell membrane, responding to α-MSH stimulation with increased cAMP production.[58] Further fluorescence microscopy and immunoblotting studies demonstrated that 4-PBA significantly decreases ubiquitination of I316S MC4R and reduces mutant-induced ER stress, suggesting that this drug functions to promote folding of mutant MC4Rs.[58] In addition, combined incubation of 4-PBA and UBE-41, a ubiquitin-activating enzyme inhibitor that directly inhibits MC4R ubiquitination, achieves more profound effects on MC4R maturation and targeting to cell membrane than treatment with either molecule alone.[58] This additive effect suggested that 4-PBA and UBE-41 target at different steps to promote MC4R exit from ER, with PBA improving receptor folding and E1 inhibitors decreasing ER-associated degradation of

MC4R. Therefore, strategies aiming at folding and/or ubiquitination step of Class II MC4Rs by using chemical chaperone and/or inhibitors of the ubiquitination machinery might render therapeutic potential for personalized treatment of specific forms of obesity.

4.3 Pharmacological chaperones for Class II mutants

Pharmacological chaperones (PCs) are small hydrophobic molecules that enter the cell, specifically bind to, and promote folding and routing of misfolded proteins by serving as a folding template and stabilizing their native intramolecular structure and/or quaternary structure (reviewed in Refs. 27,30). Currently, the most common strategies to identify PCs are through screening of small molecule libraries and optimization of substrates, inhibitors of enzymes, or ligands of receptors and ion channels.

Compared to chemical chaperones, PCs with high specificity always achieve higher efficacy in rescuing a broader spectrum of Class II mutants.[30] Indeed, PC therapy for the treatment of protein conformational diseases has been considered as a promising alternative to gene therapy and enzyme replacement therapy. To date, PC therapy has been investigated in multiple diseases, including long QT syndrome caused by misfolded HERG potassium channel,[60] cystic fibrosis caused by misfolded cystic fibrosis transmembrane conductance regulator,[61,62] Fabry disease caused by misfolded α-galactosidase A,[63] Gaucher disease caused by misfolded β-glucosidase,[64] aspartylglucosaminuria caused by misfolded aspartylglucosaminidase,[65] and endocrine diseases caused by misfolded GPCRs (reviewed in Refs. 27,30).

4.3.1 Rhodopsin

The pharmacological chaperoning activities of retinoids (9-*cis* and 11-*cis*-retinal) and their analogs (11-*cis*-7-ring retinal) have been reported by several groups, who demonstrated that 9- and 11-*cis*-retinal as well as 11-*cis*-7-ring retinal improve the accumulation of correctly folded T17M and P23H rhodopsins in cell culture.[66–69] In addition, 11-*cis*-retinal was shown to have a general effect in restoring the intracellularly retained rhodopsin mutants since the folding and trafficking of Class II rhodopsins, including T4K, N15S, V20G, P23A/L, and Q28H, are also promoted by this compound.[69,70]

In a T17M transgenic mouse model of retinitis pigmentosa, vitamin A supplementation slows the course of photoreceptor degeneration by preserving both ERG a- and b-wave amplitudes and retinal morphology.[66] Further in vitro study showed that T17M rhodopsin instability can indeed be alleviated to a large extent through 11-*cis*-retinal binding,[66] suggesting that the beneficial effect of vitamin A supplementation on the T17M mice might be due to increased supply of 11-*cis*-retinal in the retina. Therefore, this study provides explanations for the therapeutic effects of vitamin A supplementation on the course of photoreceptor degeneration in patients with retinitis pigmentosa.

4.3.2 V2R

In the GPCRs, the first attempt to utilize PCs to rescue misfolded *AVPR2* mutations pathogenic for NDI was reported by Bouvier and colleagues, who showed that two cell-permeant V2R antagonists, SR121463 (satavaptan) and VPA-985 (lixivaptan), restore both cell surface targeting and agonist-stimulated signaling for 8 of 15 Class II V2R mutants through binding to the intracellular precursor receptors to promote its proper folding and maturation.[71] Subsequently, an increasing number of PCs for V2R were discovered. These include antagonists such as SR49059 (relcovaptan),[72,73] YM087 (conivaptan),[74] OPC31260 (mozavaptan),[75] and OPC41061 (tolvaptan),[75] agonists such as OPC51803,[76] VA88,[76] and VA89,[76] as well as biased agonists such as MCF14,[77] MCF18,[77] and MCF57.[77]

It should be noted that investigating the efficacy of V2R PCs in vivo is very difficult due to lack of transgenic animal models. Male mice with complete loss of V2R die within the first week after birth due to hypernatremic dehydration.[78] The only available mouse model for NDI study is a mouse carrying E242X *AVPR2*, a mutation that has not been found in humans yet.[79] Therefore, it is of urgent need to generate V2R knock-in animals for assessing the efficacy and safety of V2R PCs in vivo.

To date, only one clinical trial has been conducted to test the efficacy of cell-permeable antagonists, in which five X-linked NDI patients carrying intracellularly retained *AVPR2* mutations (R137H, W164S, and Δ62–64) were recruited.[71] Short-term treatment with SR49059 significantly decreases 24-h urinary output and water intake and increases urine osmolality in all subjects without significant change in sodium, potassium, and creatinine excretions.[71] However, this clinical trial was discontinued due to side effects possibly associated with interference with the cytochrome P450 metabolic pathway.[80] Recently, V2R antagonist OPC41061 (Tolvaptan), a PC for *AVPR2* mutations with high efficacy at clinically feasible concentrations, was approved in the United States and Europe to treat hyponatremia and heart failure.[81,82] The high rescuing ability and positive safety profile make OPC41061 a promising candidate to be tested for the treatment of X-linked NDI.

4.3.3 MC4R

In 2009, we identified the first PC for the MC4R. Using stably transfected HEK293 cells, we showed that ML00253764, a small molecule MC4R antagonist and partial inverse agonist,[83,84] restores the cell surface expression and function of two Class II MC4R mutants, C84R and W174C.[85] Similar rescuing effects of this molecule are also observed in several additional intracellularly retained MC4R mutants.[16,86]

Since then, we and others identified 11 additional MC4R PCs, including MTHP, PPPone, MPCI, DCPMP, NBP, Ipsen 5i, Ipsen 17, and 4 MC4R-selective antagonists without disclosed identities, that rescue the cell surface targeting and signaling of different misfolded MC4Rs with distinct potency and efficacy profiles (Table 1[87–90]). Among these antagonists, Ipsen 17 (0.96 nM) and Ipsen 5i (2.0 nM) with the highest affinities exert extremely broad rescue spectra, restoring the cell surface expression of 11 of 12 and 21 of 23 selected misfolded MC4Rs, respectively,[87,88] probably because Ipsen 17 has a rescue EC_{50} around 10 nM and Ipsen 5i can rescue mutant MC4Rs at a concentration as low as 1 nM.[87,88] Other antagonists, including MTHP (K_i 0.09 μM), MPCI (K_i 0.22 μM), and PPPone (K_i 0.69 μM), have less broad rescue spectra and efficiency.[89] These observations are consistent with the concept that a higher binding affinity exerted by the compound usually correlates with higher chaperoning activity.

On the other hand, the higher affinity of the antagonist often hinders subsequent dissociation from the receptor and displacement by the endogenous ligands. Indeed, the misfolded MC4Rs rescued to the cell membrane by NBP fail to respond to NDP-MSH in generating cAMP, likely due to prolonged occupation at the receptors.[89] However, the majority of mutants, once rescued to the cell membrane by Ipsen 17 and Ipsen 5i, respond to the stimulation of NDP- or α-MSH with significantly increased cAMP production compared to the untreated controls,[87,88] which might result from relatively low functional antagonist potencies (77 and 44 nM for Ipsen 17 and Ipsen 5i, respectively) and rapid dissociation rate from the receptor.[87,88]

In addition to antagonists, we also identified a selective MC4R agonist, THIQ,[92] acting as a PC for misfolded receptors. In our study, THIQ is shown to restore the cell surface expression and signaling of 7 of 10 mutant MC4Rs in physiologically relevant neuronal cell lines with mutant-dependent rescue EC_{50}s ranging from 10 nM to 1 mM.[91] However, in the mammalian HEK293 cells, THIQ only rescues 3 of 10 mutants,[91] suggesting that the chaperoning activity conferred by THIQ is cell line dependent, with broader

Table 1 Cellular trafficking and signaling rescue of misfolded obesity-associated MC4R mutants by PCs.

Compound	Nomenclature	K_i (nM)	Cell surface-rescued mutants	Signaling-rescued mutants	Cell lines	References
ML00253764	2-[2-[2-(5-Bromo-2-methoxyphenyl)-ethyl]-3-fluorophenyl]-4,5-dihydro-1H-imidazol	160	S58C, N62S, I69T, C84R, G98R, I102S, Y157S, M161T, T162I, R165Q, R165W, W174C, I195S, I269N, C271Y, A303P, I316S	N62S, I69T, P78L, C84R, Y157S, M161T, T162I, R165W, W174C, C271Y	HEK293	16,85
			N62S, C84R, W174C	N62S, C84R, W174C, C271Y	Neuro2a and N1E-115	Unpublished data
Ipsen 5i	NA	2.0	S58C, N62S, I69T, P78L, C84R, G98R, I102S, Y157S, M161T, T162I, R165Q, R165W, W174C, I195S, P260Q, I269N, C271Y, P299H, A303P, I316S	N62S, I69R, I69T, P78L, C84R, Y157S, M161T, T162I, R165Q, R165W, P260Q, C271Y, A303P	HEK293	16,87
			N62S, I69R, P78L, C84R, G98R, W174C, P260Q, C271Y	N62S, I69R, P78L, C84R, W174C, P260Q, C271Y	Neuro2a and N1E-115	87
Ipsen 17	NA	0.96	S58C, E61K, N62S, I69T, P78L, C84R, G98R, T162I, R165W, W174C, C271Y	S58C, E61K, N62S, I69T, P78L, C84R, G98R, T162I, R165W, W174C, C271Y	HEK293	88
MTHP	2-(2-(2-Methoxy-5-itrobenzylthio)-phenyl)-1,4,5,6-tetrahydropyrimidine	94	S58C, I125K, R165Q, R165W	I69T, R165Q, R165W	HEK293T	89
PPPone	3-(4-(2-(4-Fluorophenyl)-2-(4-methylpiperazin-1-yl)ethyl)piperazin-1-yl)-2-methyl-1-phenylpropan-1-one	691	S58C, E61K, N62S, I69T, T162I, R165Q, R165W, C271Y	N62S, I69T, R165Q, R165W	HEK293T	89

MPCI	2-(5-Bromo-2-methoxyphenethyl)-*N*-(*N*-((1-ethylpiperidin-4-yl)methyl) carbamimidoyl)-3-fluorobenzamide	218	T162I, R165Q, R165W, C271Y	R165Q, R165W	HEK293T	89
DCPMP	*N*-((2*R*)-3(2,4-Dichlorophenyl)-1-(4-(2-((1-methoxypropan-2-ylamino) methyl)phenyl)piperazin-1-yl)-1-oxopropan-2-yl)propionamide	23.7	S58C, E61K, N62S, I69T, I125K, T162I, R165Q, R165W, C271Y	S58C, E61K, N62S, I69T, T162I, R165Q, R165W, C271Y	HEK293T	89
NBP	1-(1-(4-Fluorophenyl)-2-(4-(4-(naphthalene-1-yl)butyl)piperazin-1-yl)ethyl)-4-methylpiperazine	2.4	S58C, E61K, N62S, I69T, I125K, T162I, R165Q, R165W, C271Y	R165Q, R165W	HEK293T	89
Compound A	NA	NA[a]	V50M, S58C, I137T	V50M, S58C	HEK293	90
Compound B	NA	NA[a]	V50M, S58C, I137T	V50M, S58C	HEK293	90
Compound C	NA	NA[a]	V50M, S58C, I137T	V50M, S58C	HEK293	90
Compound D	NA	NA[a]	V50M, S58C, I137T	V50M, S58C	HEK293	90
THIQ	*N*-[(3*R*)-1,2,3,4-Tetrahydroisoquinolinium-3-ylcarbonyl]-(1*R*)-1-(4-chlorobenzyl)-2-[4-cyclohexyl-4-(1H-1,2,4-triazol-1-ylmethyl) piperidin-1-yl]-2-oxoethylamine	1.2	N62S, C84R, C271Y	N62S, P78L, C84R, P260Q	HEK293	91
			N62S, I69R, P78L, C84R, W174C, P260Q, C271Y	N62S, I69R, P78L, C84R, W174C, P260Q, C271Y	Neuro2a and N1E-115	91

[a]The binding affinity of the drug for the WT MC4R.
NA, data not avaliable.

rescue spectra and higher rescue efficacy in neuronal cells than in HEK293 cells. THIQ exerts several advantages to be utilized as a therapeutic option for the treatment of obesity caused by misfolded MC4R: (1) THIQ does not need to be displaced by the endogenous ligands in order to activate the receptor; (2) THIQ has good oral bioavailability[93]; and (3) THIQ can cross the brain blood barrier and reach CNS.[93] However, prolonged treatment with THIQ also promotes receptor desensitization and internalization and/or interferes with recycling of internalized receptor, as evidenced by dramatic decrease in the cell membrane expression and signaling of WT MC4R upon 24-h incubation with THIQ.[91] In this scenario, screening for allosteric modulators without blocking endogenous ligand binding and inducing receptor downregulation exerts better therapeutic potential and is favored in future studies.

4.3.4 GnRHR

The first PC for the GnRHR was identified by Conn's lab. Using a heterologous expression system, a GnRHR peptidomimetic antagonist belonging to indole class was shown to rescue almost all five naturally occurring and eight artificially lab-generated misfolded GnRHR mutants, as measured by radioligand binding or buserelin-induced IP production.[94] Subsequently, more peptidomimetic antagonists developed from four different chemical classes, including indoles, quinolones, thienopyrimidinediones, and erythromycin-derived macrolides, have been demonstrated to have chaperoning activities (Table 2). Of 17 misfolded GnRHR mutants tested, 14 are partially or completely rescued by these compounds.[95,96,98–101] Similar to results observed in the V2R and MC4R, the rescue efficacy correlates with the binding affinity for the WT receptor. Indeed, among all the chemicals tested, the compounds with higher affinity, IN30 (IC_{50} 0.2 nM), followed by Q89 (IC_{50} 0.3 nM), IN3 (IC_{50} 0.6 nM), A177775 (IC_{50} 17.7 nM), and A2222509 (IC_{50} 20 nM), are shown to be the most effective in restoring the function of misfolded GnRHR mutants.[96,102]

These molecules are potentially desirable PCs for GnRHR since they are predicted to have the ability to permeate the cell membrane, interact with the GnRHR with high specificity and variable affinity, and bind reversibly to the target mutants.[102] Despite some advances made in revealing GnRHR structure and drug interactions, the exact mechanism for PC action on GnRHR is still uncertain except for mutant E90K, the first inactivating GnRHR mutant with defective receptor expression. This mutation inhibits the formation of the E90-K121 salt bridge and destabilizes the interaction

Table 2 Signaling rescue of misfolded hypergonadotropic hypogonadism-associated GnRHR mutants by PCs.

Compound	Nomenclature	IC_{50}[a] (nM)	Signaling-rescued mutants	Cell lines	References
IN3	(2*S*)-2-[5-[2-(2-Azabicyclo[2.2.2]oct-2-yl)-1,1-dimethyl-2-oxoethyl]-2-(3,5-dimethylphenyl)-1H-indol-3-yl]-N-(2-pyridin-4-ylethyl)propan-1-amine	0.6	N10K, T32I, E90K, Q106R, A129D, R139H, C200Y, R262Q, L266R, C279Y, Y284C, A171T/Q106R, R262Q/Q106R, R262Q/A129D, R262Q/Y284C, R262Q/L314X	COS-7	95–97
IN30	(2*S*)-2-[5-[2-(2-Azabicyclo[2.2.2]oct-2-yl)-1,1-dimethyl-2-oxoethyl]-2-(3,5-dimethylphenyl)-1H-indol-3-yl]-*N*-{2-[4-(methylsulfinyl)phenyl]ethyl}propan-1-amine	0.2	N10K, T32I, E90K, Q106R, R139H, C200Y, R262Q, C279Y, Y284C	COS-7	96
IN31b	(2*S*)-*N*-[2-(4-Carboxyphenyl)ethyl]-2-[5-[1,1-dimethyl-2-(4-methylpiperazin-1-yl)-2-oxoethyl]-2-(3,5-dimethylphenyl)-1H-indol-3-yl]propan-1-aminium trifluoroacetate	3.4	N10K, T32I, R262Q, Y284C	COS-7	96
Q89	7-Chloro-2-oxo-4-{2-[(2*S*)-piperidin-2-yl]ethoxy}-*N*-pyrimidin-4-yl-3-(3,4,5-trimethylphenyl)-1,2-dihydroquinoline-6-carboxamide	0.3	N10K, T32I, E90K, Q106R, C200Y, R262Q, L266R, Y284C	COS-7	96
Q76	*N*-(7-Chloro-3-(3,5-dimethylphenyl)-2-oxo-4-{2-[(2*S*)-piperidin-2-yl]ethoxy}-1,2-dihydroquinolin-6-yl)-*N′*-cyclopropylurea	2.3	N10K, T32I, R262Q, Y284C	COS-7	96

Continued

Table 2 Signaling rescue of misfolded hypergonadotropic hypogonadism-associated GnRHR mutants by PCs.—cont'd

Compound	Nomenclature	IC_{50} (nM)	Signaling-rescued mutants	Cell lines	References
Q08	(2*S*)-2-(2-{[7-Chloro-6-[(6,7-dimethoxy-3,4-dihydroisoquinolin-2(1H)-yl)carbonyl]-3-(3,5-dimethylphenyl)-2-oxo-1,2-dihydroquinolin-4-yl]oxy}ethyl) piperidinium trifluoroacetate	186	N10K, T32I, R262Q, Y284C	COS-7	96
A-177775	3′-*N*-Desmethyl-3′-*N*-cyclopentyl-11-deoxy-11-[carboxy-(3,4-dichlorophenylethylamino)]-6-*O*-methyl-erythromycin A 11,12-(cyclic carbamate)	17.7	N10K, T32I, E90K, Q106R, C200Y, R262Q, Y284C, Des325, DesL237-L241, C278A	COS-7	96
A-222509	3′,3′-*N*-Desmethyl-3′,3′-*N*-cyclopropylmethyl-11-deoxy-11-[carboxy-(3-chloro,4-fluoro-phenylethylamino)]-6-*O*-methyl-erythromycin A 11,12-(cyclic carbamate)	20	N10K, T32I, E90K, Q106R, C200Y, R262Q, Y284C, Des325, DesL237-L241, C278A	COS-7	96
A-64755	11-Deoxy-11-[carboxy-phenylethylamino]-6-*O*-methyl-erythromycin A 11,12-(cyclic carbamate)	669.5	N10K, T32I, Q106R, C200Y, R262Q, Y284C, C278A	COS-7	96

[a]The efficacy of each of the drugs tested on the rescue of a single mutant, E90K, as assessed by inositol phosphate (IP) production.

between TM2 and TM3.[103] Taking advantage of computer modeling, several groups showed that indoles, quinolones, and erythromycin macrolides can stabilize the conformation of mutant E90K by forming a bridge between residues D98 located at extracellular proportion of TM1 and K121 located at TM3, which acts as surrogate for naturally existing E90-K121 salt bridge to stabilize the orientation of the TMs in the mutant receptor.[103–105]

In order to test the therapeutic utility of PCs in vivo, Conn and colleagues generated a *GNRHR* E90K knock-in mouse model manifesting with hypogonadotropic hypogonadism, similar to symptoms observed in human carriers.[106] With this mouse model, pulsatile delivery of IN3 to anterior pituitary was shown to promote the anterograde trafficking of E90K GnRHR to the gonadotroph cell membrane, increase GnRHR activity as assayed by LH release, as well as rescue testicular weight, testosterone production, and spermatogenesis.[106]

4.3.5 Gonadotropin receptors

When considerable efforts were devoted to search for orally active non-peptide gonadotropin analogs with therapeutic potentials, thienopyrimidines emerge as the most advanced series of compounds. Among them are the cell-permeant small molecule agonists, Org 43553 and its trifluoracetic acid form Org 42599, which display high potency and efficacy at the LHCGR but relatively lower potency at the FSHR.[107–109] The chaperoning activity of Org 42599 was first investigated by Newton et al., who demonstrated that this compound rescues the folding and cell surface localization, as well as signaling, of misfolded A593P and S616Y LHCGRs that cause Leydig cell hypoplasia in man.[110] The rescue effect of Org 42599 seems to be mutant specific since treatment of this compound on 20 distinct misfolded LHCGR mutants results in a spectrum of responses from no response to a full restoration in cell surface expression levels.[109] Interestingly, Org 42599 acts in an allosteric manner since it does not displace ^{125}I-labeled human LH binding to the WT receptor.[110] Therefore, the binding of Org 42599 to mutant LHCGRs will not interfere with the subsequent receptor activation by endogenous ligands. In addition, Newton and colleagues also demonstrated that Org 42599 can act as a PC for FSHR, rescuing the cell surface expression and activity of misfolded naturally occurring and laboratory-generated FSHR mutations.[109] Given the good oral bioavailability, safety, and tolerability, Org 32599 might represent a good candidate for treating reproductive diseases caused by misfolded mutations in *LHCGR* or *FSHR*. However, selectivity is a concern.

4.4 Novel ligands for Class III and IV mutants

It has been shown that several GPCR mutants are defective in binding or signaling for one ligand but not others,[111,112] indicating that Class III and IV mutants could bind and be activated by more potent or allosteric exogenous ligands. Searching for novel ligands with the ability to rescue the binding- or signaling-defective mutants might represent a therapeutic option for correcting these defects.

4.4.1 MC4R

In a comprehensive study of the functional properties of naturally occurring *MC4R* mutations, Yeo et al. showed that I316S has reduced affinity for α-MSH but retains normal affinity for the antagonist AgRP, indicating a novel molecular mechanism of receptor dysfunction whereby mutation impairs the affinity of a receptor for its natural agonist vs antagonist.[112]

Consistent with these observations, Xiang et al. performed several comprehensive studies to search for drugs with the ability to correct the signaling of mutant MC4Rs that have decreased agonist potency of at least one endogenous melanocortin. By screening from the synthetic MC4R agonist libraries, they demonstrated that NDP-MSH, MTII, THIQ, and AMW3-130, but not JRH887-9, JRH420-12, JRH322-18, JB25, AMW3-106, restore normal functional response of 12 signaling-defective mutant MC4Rs with decreased potency for α-, β-, γ_2-MSH and ACTH (1–24).[113–115] It is well recognized that structurally different ligands can induce distinct downstream signaling by stabilizing the receptor in different conformations. The different cAMP responses induced by NDP-MSH, MTII, THIQ, and AMW3-130 and that by endogenous agonists and some synthetic agonists mentioned earlier might be due to subtle differences of the MC4R conformations stabilized by these agonists. NDP-MSH, MTII, THIQ, and AMW3-130 but not others can stabilize these signaling-defective mutants in conformations that are similar to that of activated WT hMC4R.

4.4.2 LHCGR and FSHR

The first clinical trial utilizing novel ligand to restore signaling of Class IV mutant was successfully performed in a Leydig cell hypoplasia type II patient carrying an *LHCGR* mutation lacking exon 10. In vitro studies showed that human LHCGR lacking exon 10 has decreased cell surface expression and fails to respond to LH stimulation.[116] Therefore, despite high serum LH levels, the patient has insufficient LHCGR activation and low testosterone levels, manifesting with a normal male external genitalia (indicating relatively normal hCG action during fetal development) but lacking pubertal development and hypogonadism (defective LH action during puberty).[117] Further hCG treatment for the patient restores the testosterone levels, testicular volume, sperm concentration, and total sperm counts at ejaculate to normal range.[117] In addition to this unique mutant, high doses of hCG can increase cAMP or IP production of the C133R LHCGR, a mutant with partially impaired receptor function.[118] These results suggest that hCG and LH might exert a possible dual mechanism of hormone binding and signal transduction, and hCG might have a general effect in rescuing Class III and IV LHCGR mutants.

During the past two decades, high-throughput screening campaigns using cell-based assays have yielded multiple allosteric gonadotropin receptor agonists with therapeutic potential. These molecules include thienopyrimidines and phenylpyrazoles for LHCGR, and thiazolidinones and hexahydroquinolines for FSHR. van Straten et al. reported the first orally active small molecule thienopyrimidine agonist, Org 41841, for the LHCGR with EC_{50} of 20 nM in the CHO-LHCGR assay.[119] This compound is able to induce testosterone production in cultured mouse Leydig cells and ovulation in female mice.[119] In addition, a series of phenylpyrazoles with mixed LH/FSH activity were reported. One of phenylpyrazoles is compound 5, which was shown to have an EC_{50} of 20 nM and efficacy of 53% in CHO-LHCGR, and an EC_{50} of 130 nM and efficacy of 73% in CHO-FSHR assays.[120] Further in vivo studies showed that compound 5 induces testosterone production in rats, though its effect on follicular development remains to be investigated.[120]

With EC_{50} of ~20 nM in CHO-FSHR cells, some thiazolidinone compounds are shown to induce estradiol production in functional rat granulosa cells through a binding site independent of the natural hormone binding site.[121] Moreover, a series of hexahydroquinolines with nanomolar activity in transfected CHO cells, such as Org 214444-0, have also been reported. Org 214444-0 was shown to behave as an allosteric agonist to stimulate rat granulosa cells with a binding affinity that is 6.5-fold higher than FSH, resulting in increased follicular development and ovulation in mature rats.[122]

Considering the distinct way of action, high potency, and relatively good oral activity, a promising therapeutic potential of these allosteric gonadotropin analogs could be envisioned for the treatment of reproductive disease caused by Class III and IV mutations in the *LHCGR* or *FSHR* genes. Further in vitro and in vivo studies are needed to confirm this hypothesis.

4.5 Gene therapies for *rhodopsin* mutations

adRP offers particular promise as targets of gene therapy due to the involvement of multiple genes, availability of transgenic animal models, and easy accessibility for experimentation. Delivery of adeno-associated virus (AAV) vector -mediated ribozymes, small RNA molecules that cleave mutant transcripts without affecting WT transcripts, to photoreceptors has been shown to delay the progression of photoreceptor degeneration in P23H rhodopsin transgenic rats through reducing the mutant transcripts.[123] The ribozyme-directed cleavage of mutant mRNAs was shown to be effective for long term (at least 8 months) and at late stages of the disease (with significant photoreceptor cell loss already occurred).[124] However, the disadvantages of this method, such as difficulties to design ribozymes, mutant-specific efficacy, and adaption of animals to ribozyme cleavage, hinder its application in clinical trials.

However, inhibition of rhodopsin with an antisense oligonucleotide that specifically targets the mutant allele in the P23H transgenic rats slows photoreceptor cell degeneration and preserves the function of photoreceptor cells in eyes.[125] Moreover, with the use of AAV for delivery, allele-independent RNA interference that suppresses ~90% mutant *RHO* in conjunction with supplementation of resistant endogenous *RHO* provides ERG and histology benefits for adRP mouse models harboring P23H or P347S *RHO*,[126,127] indicating that AAV-delivered suppression and replacement may well provide future therapies for patients with *RHO*-linked adRP.

The bacterial CRISPR/Cas system has proven to be a simplified, versatile, and efficient tool for genetic editing in various organisms and cell types. Although the application of CRISPR/Cas9 in vivo is rarely reported, encouraging results have been obtained in several studies investigating its efficacy in treating *RHO*-associated adRP.

In 2016, Latella et al. reported the first CRISPR–Cas9-mediated manipulation of the human *RHO* gene in the transgenic P23H mouse model.[128] In their study, delivery of a plasmid encoding for a Cas9 nuclease electroporated with a guide RNA targeting the P23H allele results in an almost complete clearance of *RHO* expression in vitro and a strong reduction of human rhodopsin in vivo.[128] Further analysis demonstrated that the designed CRISPR/Cas9 plasmids have minimal cytotoxicity, as evidenced by no induced apoptosis in CRISPR/Cas9-expressing cells.[128] However, despite a high electroporation efficiency in transfected HeLa cells (up to ~33% positive editing), a lower knock-out efficiency in mouse retinas was observed, likely due to the inefficient delivery of the plasmids and/or an already reported incomparable genome editing efficiency in vitro vs in vivo.[128] Similarly, delivery of the mutant-specific CRISPR guides was reported to provide positive outcomes in patient-derived induced pluripotent stem cells and pig retina carrying P23H *RHO* by creating a frameshift and premature stop that could prevent transcription of the disease-causing mutant.[129]

In addition to directly targeting the *RHO* gene, disruption of *Nrl*, a gene encoding for neural retina-specific leucine zipper protein that determines rod fate during photoreceptor development, by administration of AAV-mediated CRISPR/Cas9 to postmitotic photoreceptors in the *Rho* knock-out and P347S *Rho* mice has also been reported to lead to improvement of rod survival and preservation of cone function.[130]

5. Summary

The last several decades have witnessed enormous progress made in the understanding of etiologies and molecular basis of multiple endocrine diseases associated with GPCR mutations. Revealing the functional consequences of different types of mutations, especially loss-of-function mutations, is essential for developing patient-tailored strategies targeting specific functional defect. The therapeutic strategies, including aminoglycosides for Class I mutants, chemical and pharmacological chaperones for Class II mutants, novel ligands for Class III and IV mutants, as well as gene therapies for adRP, have been investigated in heterologous cell lines, animal models, and/or clinical trials. These studies not only contribute to a better understanding of GPCR biology but also provide insights in more efficient personalized medicine.

References

1. Hauser AS, Attwood MM, Rask-Andersen M, Schioth HB, Gloriam DE. Trends in GPCR drug discovery: new agents, targets and indications. *Nat Rev Drug Discov.* 2017;16:829–842.
2. Dryja TP, McGee TL, Reichel E, et al. A point mutation of the rhodopsin gene in one form of retinitis pigmentosa. *Nature.* 1990;343:364–366.
3. Schoneberg T, Schulz A, Biebermann H, Hermsdorf T, Rompler H, Sangkuhl K. Mutant G-protein-coupled receptors as a cause of human diseases. *Pharmacol Ther.* 2004;104:173–206.
4. Tao YX. Inactivating mutations of G protein-coupled receptors and diseases: structure-function insights and therapeutic implications. *Pharmacol Ther.* 2006;111:949–973.

5. Palczewski K, Kumasaka T, Hori T, et al. Crystal structure of rhodopsin: a G protein-coupled receptor. *Science*. 2000;289:739–745.
6. Smith SO. Structure and activation of the visual pigment rhodopsin. *Annu Rev Biophys*. 2010;39:309–328.
7. Sung CH, Schneider BG, Agarwal N, Papermaster DS, Nathans J. Functional heterogeneity of mutant rhodopsins responsible for autosomal dominant retinitis pigmentosa. *Proc Natl Acad Sci USA*. 1991;88:8840–8844.
8. Sung CH, Tai AW. Rhodopsin trafficking and its role in retinal dystrophies. *Int Rev Cytol*. 2000;195:215–267.
9. Bichet DG. V2R mutations and nephrogenic diabetes insipidus. *Prog Mol Biol Transl Sci*. 2009;89:15–29.
10. Moeller HB, Rittig S, Fenton RA. Nephrogenic diabetes insipidus: essential insights into the molecular background and potential therapies for treatment. *Endocr Rev*. 2013;34:278–301.
11. Vaisse C, Clement K, Guy-Grand B, Froguel P. A frameshift mutation in human MC4R is associated with a dominant form of obesity. *Nat Genet*. 1998;20:113–114.
12. Yeo GS, Farooqi IS, Aminian S, Halsall DJ, Stanhope RG, O'Rahilly S. A frameshift mutation in MC4R associated with dominantly inherited human obesity. *Nat Genet*. 1998;20:111–112.
13. Hinney A, Volckmar AL, Knoll N. Melanocortin-4 receptor in energy homeostasis and obesity pathogenesis. *Prog Mol Biol Transl Sci*. 2013;114:147–191.
14. Tao YX. Mutations in melanocortin-4 receptor and human obesity. *Prog Mol Biol Transl Sci*. 2009;88:173–204.
15. Farooqi IS, Keogh JM, Yeo GS, Lank EJ, Cheetham T, O'Rahilly S. Clinical spectrum of obesity and mutations in the melanocortin 4 receptor gene. *N Engl J Med*. 2003;348:1085–1095.
16. Huang H, Wang W, Tao YX. Pharmacological chaperones for the misfolded melanocortin-4 receptor associated with human obesity. *Biochim Biophys Acta*. 2017;1863:2496–2507.
17. de Roux N, Young J, Misrahi M, et al. A family with hypogonadotropic hypogonadism and mutations in the gonadotropin-releasing hormone receptor. *N Engl J Med*. 1997;337:1597–1602.
18. Chevrier L, Guimiot F, de Roux N. GnRH receptor mutations in isolated gonadotropic deficiency. *Mol Cell Endocrinol*. 2011;346:21–28.
19. Ulloa-Aguirre A, Zarinan T, Dias JA, Conn PM. Mutations in G protein-coupled receptors that impact receptor trafficking and reproductive function. *Mol Cell Endocrinol*. 2014;382:411–423.
20. Pierce JG, Parsons TF. Glycoprotein hormones: structure and function. *Annu Rev Biochem*. 1981;50:465–495.
21. Ascoli M, Fanelli F, Segaloff DL. The lutropin/choriogonadotropin receptor, a 2002 perspective. *Endocr Rev*. 2002;23:141–174.
22. Meduri G, Bachelot A, Cocca MP, et al. Molecular pathology of the FSH receptor: new insights into FSH physiology. *Mol Cell Endocrinol*. 2008;282:130–142.
23. Latronico AC, Anasti J, Arnhold IJ, et al. Brief report: testicular and ovarian resistance to luteinizing hormone caused by inactivating mutations of the luteinizing hormone-receptor gene. *N Engl J Med*. 1996;334:507–512.
24. Desai SS, Roy BS, Mahale SD. Mutations and polymorphisms in FSH receptor: functional implications in human reproduction. *Reproduction*. 2013;146:R235–R248.
25. Newton CL, Anderson RC, Katz AA, Millar RP. Loss-of-function mutations in the human luteinizing hormone receptor predominantly cause intracellular retention. *Endocrinology*. 2016;157:4364–4377.

26. Conn PM, Ulloa-Aguirre A, Ito J, Janovick JA. G protein-coupled receptor trafficking in health and disease: lessons learned to prepare for therapeutic mutant rescue in vivo. *Pharmacol Rev*. 2007;59:225–250.
27. Tao YX, Conn PM. Chaperoning G protein-coupled receptors: from cell biology to therapeutics. *Endocr Rev*. 2014;35:602–647.
28. Tao YX, Segaloff DL. Functional characterization of melanocortin-4 receptor mutations associated with childhood obesity. *Endocrinology*. 2003;144:4544–4551.
29. Tao YX. Molecular mechanisms of the neural melanocortin receptor dysfunction in severe early onset obesity. *Mol Cell Endocrinol*. 2005;239:1–14.
30. Tao YX, Conn PM. Pharmacoperones as novel therapeutics for diverse protein conformational diseases. *Physiol Rev*. 2018;98:697–725.
31. Wilschanski M, Yahav Y, Yaacov Y, et al. Gentamicin-induced correction of CFTR function in patients with cystic fibrosis and CFTR stop mutations. *N Engl J Med*. 2003;349:1433–1441.
32. Barton-Davis ER, Cordier L, Shoturma DI, Leland SE, Sweeney HL. Aminoglycoside antibiotics restore dystrophin function to skeletal muscles of mdx mice. *J Clin Invest*. 1999;104:375–381.
33. Lai CH, Chun HH, Nahas SA, et al. Correction of ATM gene function by aminoglycoside-induced read-through of premature termination codons. *Proc Natl Acad Sci USA*. 2004;101:15676–15681.
34. James PD, Raut S, Rivard GE, et al. Aminoglycoside suppression of nonsense mutations in severe hemophilia. *Blood*. 2005;106:3043–3048.
35. Mort M, Ivanov D, Cooper DN, Chuzhanova NA. A meta-analysis of nonsense mutations causing human genetic disease. *Hum Mutat*. 2008;29:1037–1047.
36. Keeling KM, Xue X, Gunn G, Bedwell DM. Therapeutics based on stop codon readthrough. *Annu Rev Genomics Hum Genet*. 2014;15:371–394.
37. Schulz A, Sangkuhl K, Lennert T, et al. Aminoglycoside pretreatment partially restores the function of truncated V(2) vasopressin receptors found in patients with nephrogenic diabetes insipidus. *J Clin Endocrinol Metab*. 2002;87:5247–5257.
38. Sangkuhl K, Schulz A, Rompler H, Yun J, Wess J, Schoneberg T. Aminoglycoside-mediated rescue of a disease-causing nonsense mutation in the V2 vasopressin receptor gene in vitro and in vivo. *Hum Mol Genet*. 2004;13:893–903.
39. Brumm H, Muhlhaus J, Bolze F, et al. Rescue of melanocortin 4 receptor (MC4R) nonsense mutations by aminoglycoside-mediated read-through. *Obesity (Silver Spring)*. 2012;20:1074–1081.
40. Sabbavarapu NM, Shavit M, Degani Y, Smolkin B, Belakhov V, Baasov T. Design of novel aminoglycoside derivatives with enhanced suppression of diseases-causing nonsense mutations. *ACS Med Chem Lett*. 2016;7:418–423.
41. Shulman E, Belakhov V, Wei G, et al. Designer aminoglycosides that selectively inhibit cytoplasmic rather than mitochondrial ribosomes show decreased ototoxicity: a strategy for the treatment of genetic diseases. *J Biol Chem*. 2014;289:2318–2330.
42. Xue X, Mutyam V, Tang L, et al. Synthetic aminoglycosides efficiently suppress cystic fibrosis transmembrane conductance regulator nonsense mutations and are enhanced by ivacaftor. *Am J Respir Cell Mol Biol*. 2014;50:805–816.
43. Vagenende V, Yap MG, Trout BL. Mechanisms of protein stabilization and prevention of protein aggregation by glycerol. *Biochemistry*. 2009;48:11084–11096.
44. Papp E, Csermely P. Chemical chaperones: mechanisms of action and potential use. *Handb Exp Pharmacol*. 2006;405–416.
45. Sato S, Ward CL, Krouse ME, Wine JJ, Kopito RR. Glycerol reverses the misfolding phenotype of the most common cystic fibrosis mutation. *J Biol Chem*. 1996;271:635–638.

46. Burrows JA, Willis LK, Perlmutter DH. Chemical chaperones mediate increased secretion of mutant alpha 1-antitrypsin (alpha 1-AT) Z: a potential pharmacological strategy for prevention of liver injury and emphysema in alpha 1-AT deficiency. *Proc Natl Acad Sci USA*. 2000;97:1796–1801.
47. Tamarappoo BK, Verkman AS. Defective aquaporin-2 trafficking in nephrogenic diabetes insipidus and correction by chemical chaperones. *J Clin Invest*. 1998;101: 2257–2267.
48. Datta R, Waheed A, Shah GN, Sly WS. Signal sequence mutation in autosomal dominant form of hypoparathyroidism induces apoptosis that is corrected by a chemical chaperone. *Proc Natl Acad Sci USA*. 2007;104:19989–19994.
49. Sobolewski A, Rudarakanchana N, Upton PD, et al. Failure of bone morphogenetic protein receptor trafficking in pulmonary arterial hypertension: potential for rescue. *Hum Mol Genet*. 2008;17:3180–3190.
50. Mendes HF, Cheetham ME. Pharmacological manipulation of gain-of-function and dominant-negative mechanisms in rhodopsin retinitis pigmentosa. *Hum Mol Genet*. 2008;17:3043–3054.
51. Jiang H, Xiong S, Xia X. Chemical chaperone 4-phenylbutyrate prevents endoplasmic reticulum stress induced by T17M rhodopsin. *Cell Biosci*. 2014;4:75.
52. Athanasiou D, Aguila M, Bellingham J, et al. The molecular and cellular basis of rhodopsin retinitis pigmentosa reveals potential strategies for therapy. *Prog Retin Eye Res*. 2018;62:1–23.
53. Vasireddy V, Chavali VR, Joseph VT, et al. Rescue of photoreceptor degeneration by curcumin in transgenic rats with P23H rhodopsin mutation. *PLoS One*. 2011;6: e21193.
54. Fernandez-Sanchez L, Lax P, Pinilla I, Martin-Nieto J, Cuenca N. Tauroursodeoxycholic acid prevents retinal degeneration in transgenic P23H rats. *Invest Ophthalmol Vis Sci*. 2011;52:4998–5008.
55. Fernandez-Sanchez L, Bravo-Osuna I, Lax P, et al. Controlled delivery of tauroursodeoxycholic acid from biodegradable microspheres slows retinal degeneration and vision loss in P23H rats. *PLoS One*. 2017;12:e0177998.
56. Robben JH, Sze M, Knoers NV, Deen PM. Rescue of vasopressin V2 receptor mutants by chemical chaperones: specificity and mechanism. *Mol Biol Cell*. 2006;17:379–386.
57. Cheong HI, Cho HY, Park HW, Ha IS, Choi Y. Molecular genetic study of congenital nephrogenic diabetes insipidus and rescue of mutant vasopressin V2 receptor by chemical chaperones. *Nephrology (Carlton)*. 2007;12:113–117.
58. Granell S, Mohammad S, Ramanagoudr-Bhojappa R, Baldini G. Obesity-linked variants of melanocortin-4 receptor are misfolded in the endoplasmic reticulum and can be rescued to the cell surface by a chemical chaperone. *Mol Endocrinol*. 2010;24:1805–1821.
59. Granell S, Serra-Juhe C, Martos-Moreno GA, et al. A novel melanocortin-4 receptor mutation MC4R-P272L associated with severe obesity has increased propensity to be ubiquitinated in the ER in the face of correct folding. *PLoS One*. 2012;7, e50894.
60. Zhou Z, Gong Q, January CT. Correction of defective protein trafficking of a mutant HERG potassium channel in human long QT syndrome. Pharmacological and temperature effects. *J Biol Chem*. 1999;274:31123–31126.
61. Sampson HM, Robert R, Liao J, et al. Identification of a NBD1-binding pharmacological chaperone that corrects the trafficking defect of F508del-CFTR. *Chem Biol*. 2011;18:231–242.
62. Wainwright CE, Elborn JS, Ramsey BW, et al. Lumacaftor-Ivacaftor in patients with cystic fibrosis homozygous for Phe508del CFTR. *N Engl J Med*. 2015;373:220–231.
63. Yam GH, Bosshard N, Zuber C, Steinmann B, Roth J. Pharmacological chaperone corrects lysosomal storage in Fabry disease caused by trafficking-incompetent variants. *Am J Physiol Cell Physiol*. 2006;290:C1076–C1082.

64. Lieberman RL, Wustman BA, Huertas P, et al. Structure of acid beta-glucosidase with pharmacological chaperone provides insight into Gaucher disease. *Nat Chem Biol.* 2007;3:101–107.
65. Banning A, Gulec C, Rouvinen J, Gray SJ, Tikkanen R. Identification of small molecule compounds for pharmacological chaperone therapy of aspartylglucosaminuria. *Sci Rep.* 2016;6:37583.
66. Li T, Sandberg MA, Pawlyk BS, et al. Effect of vitamin A supplementation on rhodopsin mutants threonine-17 → methionine and proline-347 → serine in transgenic mice and in cell cultures. *Proc Natl Acad Sci USA.* 1998;95:11933–11938.
67. Noorwez SM, Kuksa V, Imanishi Y, et al. Pharmacological chaperone-mediated in vivo folding and stabilization of the P23H-opsin mutant associated with autosomal dominant retinitis pigmentosa. *J Biol Chem.* 2003;278:14442–14450.
68. Noorwez SM, Malhotra R, McDowell JH, Smith KA, Krebs MP, Kaushal S. Retinoids assist the cellular folding of the autosomal dominant retinitis pigmentosa opsin mutant P23H. *J Biol Chem.* 2004;279:16278–16284.
69. Krebs MP, Holden DC, Joshi P, Clark 3rd CL, Lee AH, Kaushal S. Molecular mechanisms of rhodopsin retinitis pigmentosa and the efficacy of pharmacological rescue. *J Mol Biol.* 2010;395:1063–1078.
70. Opefi CA, South K, Reynolds CA, Smith SO, Reeves PJ. Retinitis pigmentosa mutants provide insight into the role of the N-terminal cap in rhodopsin folding, structure, and function. *J Biol Chem.* 2013;288:33912–33926.
71. Morello JP, Salahpour A, Laperriere A, et al. Pharmacological chaperones rescue cell-surface expression and function of misfolded V2 vasopressin receptor mutants. *J Clin Invest.* 2000;105:887–895.
72. Bernier V, Lagace M, Lonergan M, Arthus MF, Bichet DG, Bouvier M. Functional rescue of the constitutively internalized V2 vasopressin receptor mutant R137H by the pharmacological chaperone action of SR49059. *Mol Endocrinol.* 2004;18: 2074–2084.
73. Wuller S, Wiesner B, Loffler A, et al. Pharmacochaperones post-translationally enhance cell surface expression by increasing conformational stability of wild-type and mutant vasopressin V2 receptors. *J Biol Chem.* 2004;279:47254–47263.
74. Bernier V, Morello JP, Zarruk A, et al. Pharmacologic chaperones as a potential treatment for X-linked nephrogenic diabetes insipidus. *J Am Soc Nephrol.* 2006;17:232–243.
75. Robben JH, Sze M, Knoers NV, Deen PM. Functional rescue of vasopressin V2 receptor mutants in MDCK cells by pharmacochaperones: relevance to therapy of nephrogenic diabetes insipidus. *Am J Physiol Renal Physiol.* 2007;292:F253–F260.
76. Robben JH, Kortenoeven ML, Sze M, et al. Intracellular activation of vasopressin V2 receptor mutants in nephrogenic diabetes insipidus by nonpeptide agonists. *Proc Natl Acad Sci USA.* 2009;106:12195–12200.
77. Jean-Alphonse F, Perkovska S, Frantz MC, et al. Biased agonist pharmacochaperones of the AVP V2 receptor may treat congenital nephrogenic diabetes insipidus. *J Am Soc Nephrol.* 2009;20:2190–2203.
78. Yun J, Schoneberg T, Liu J, et al. Generation and phenotype of mice harboring a nonsense mutation in the V2 vasopressin receptor gene. *J Clin Invest.* 2000;106:1361–1371.
79. Robben JH, Deen PM. Pharmacological chaperones in nephrogenic diabetes insipidus: possibilities for clinical application. *BioDrugs.* 2007;21:157–166.
80. Los EL, Deen PM, Robben JH. Potential of nonpeptide (ant)agonists to rescue vasopressin V2 receptor mutants for the treatment of X-linked nephrogenic diabetes insipidus. *J Neuroendocrinol.* 2010;22:393–399.
81. Schrier RW, Gross P, Gheorghiade M, et al. Tolvaptan, a selective oral vasopressin V2-receptor antagonist, for hyponatremia. *N Engl J Med.* 2006;355:2099–2112.
82. Farmakis D, Filippatos G, Kremastinos DT, Gheorghiade M. Vasopressin and vasopressin antagonists in heart failure and hyponatremia. *Curr Heart Fail Rep.* 2008;5:91–96.

83. Tao YX, Huang H, Wang ZQ, Yang F, Williams JN, Nikiforovich GV. Constitutive activity of neural melanocortin receptors. *Methods Enzymol.* 2010;484:267–279.
84. Vos TJ, Caracoti A, Che JL, et al. Identification of 2-[2-[2-(5-bromo-2-methoxyphenyl)-ethyl]-3-fluorophenyl]-4,5-dihydro-1H-imidazole (ML00253764), a small molecule melanocortin 4 receptor antagonist that effectively reduces tumor-induced weight loss in a mouse model. *J Med Chem.* 2004;47:1602–1604.
85. Fan ZC, Tao YX. Functional characterization and pharmacological rescue of melanocortin-4 receptor mutations identified from obese patients. *J Cell Mol Med.* 2009;13:3268–3282.
86. Tao YX. The melanocortin-4 receptor: physiology, pharmacology, and pathophysiology. *Endocr Rev.* 2010;31:506–543.
87. Tao YX, Huang H. Ipsen 5i is a novel potent pharmacoperone for intracellularly retained melanocortin-4 receptor mutants. *Front Endocrinol (Lausanne).* 2014;5:131.
88. Wang XH, Wang HM, Zhao BL, Yu P, Fan ZC. Rescue of defective MC4R cell-surface expression and signaling by a novel pharmacoperone Ipsen 17. *J Mol Endocrinol.* 2014;53:17–29.
89. Rene P, Le Gouill C, Pogozheva ID, et al. Pharmacological chaperones restore function to MC4R mutants responsible for severe early-onset obesity. *J Pharmacol Exp Ther.* 2010;335:520–532.
90. Ward NA, Hirst S, Williams J, Findlay JB. Pharmacological chaperones increase the cell-surface expression of intracellularly retained mutants of the melanocortin 4 receptor with unique rescuing efficacy profiles. *Biochem Soc Trans.* 2012;40:717–720.
91. Huang H, Tao YX. A small molecule agonist THIQ as a novel pharmacoperone for intracellularly retained melanocortin-4 receptor mutants. *Int J Biol Sci.* 2014;10:817–824.
92. Sebhat IK, Martin WJ, Ye Z, et al. Design and pharmacology of N-[(3R)-1,2,3,4-tetrahydroisoquinolinium-3-ylcarbonyl]-(1R)-1-(4-chlorobenzyl)-2-[4-cyclohexyl-4-(1H-1,2,4-triazol-1-ylmethyl)piperidin-1-yl]-2-oxoethylamine (1), a potent, selective, melanocortin subtype-4 receptor agonist. *J Med Chem.* 2002;45:4589–4593.
93. Chen M, Cai M, Aprahamian CJ, et al. Contribution of the conserved amino acids of the melanocortin-4 receptor in [corrected] [Nle4,D-Phe7]-alpha-melanocyte-stimulating [corrected] hormone binding and signaling. *J Biol Chem.* 2007;282: 21712–21719.
94. Janovick JA, Maya-Nunez G, Conn PM. Rescue of hypogonadotropic hypogonadism-causing and manufactured GnRH receptor mutants by a specific protein-folding template: misrouted proteins as a novel disease etiology and therapeutic target. *J Clin Endocrinol Metab.* 2002;87:3255–3262.
95. Leanos-Miranda A, Janovick JA, Conn PM. Receptor-misrouting: an unexpectedly prevalent and rescuable etiology in gonadotropin-releasing hormone receptor-mediated hypogonadotropic hypogonadism. *J Clin Endocrinol Metab.* 2002;87: 4825–4828.
96. Janovick JA, Goulet M, Bush E, Greer J, Wettlaufer DG, Conn PM. Structure-activity relations of successful pharmacologic chaperones for rescue of naturally occurring and manufactured mutants of the gonadotropin-releasing hormone receptor. *J Pharmacol Exp Ther.* 2003;305:608–614.
97. Leanos-Miranda A, Ulloa-Aguirre A, Janovick JA, Conn PM. In vitro coexpression and pharmacological rescue of mutant gonadotropin-releasing hormone receptors causing hypogonadotropic hypogonadism in humans expressing compound heterozygous alleles. *J Clin Endocrinol Metab.* 2005;90:3001–3008.
98. Ashton WT, Sisco RM, Kieczykowski GR, et al. Orally bioavailable, indole-based nonpeptide GnRH receptor antagonists with high potency and functional activity. *Bioorg Med Chem Lett.* 2001;11:2597–2602.

99. Ashton WT, Sisco RM, Yang YT, et al. Substituted indole-5-carboxamides and -acetamides as potent nonpeptide GnRH receptor antagonists. *Bioorg Med Chem Lett.* 2001;11:1723–1726.
100. Ashton WT, Sisco RM, Yang YT, et al. Potent nonpeptide GnRH receptor antagonists derived from substituted indole-5-carboxamides and -acetamides bearing a pyridine side-chain terminus. *Bioorg Med Chem Lett.* 2001;11:1727–1731.
101. Ulloa-Aguirre A, Janovick JA, Leanos-Miranda A, Conn PM. Misrouted cell surface GnRH receptors as a disease aetiology for congenital isolated hypogonadotrophic hypogonadism. *Hum Reprod Update.* 2004;10:177–192.
102. Conn PM, Ulloa-Aguirre A. Pharmacological chaperones for misfolded gonadotropin-releasing hormone receptors. *Adv Pharmacol.* 2011;62:109–141.
103. Janovick JA, Patny A, Mosley R, et al. Molecular mechanism of action of pharmacoperone rescue of misrouted GPCR mutants: the GnRH receptor. *Mol Endocrinol.* 2009;23:157–168.
104. Conn PM, Janovick JA. Trafficking and quality control of the gonadotropin releasing hormone receptor in health and disease. *Mol Cell Endocrinol.* 2009;299:137–145.
105. Jardon-Valadez E, Ulloa-Aguirre A, Pineiro A. Modeling and molecular dynamics simulation of the human gonadotropin-releasing hormone receptor in a lipid bilayer. *J Phys Chem B.* 2008;112:10704–10713.
106. Janovick JA, Stewart MD, Jacob D, et al. Restoration of testis function in hypogonadotropic hypogonadal mice harboring a misfolded GnRHR mutant by pharmacoperone drug therapy. *Proc Natl Acad Sci USA.* 2013;110:21030–21035.
107. van Koppen CJ, Zaman GJ, Timmers CM, et al. A signaling-selective, nanomolar potent allosteric low molecular weight agonist for the human luteinizing hormone receptor. *Naunyn Schmiedebergs Arch Pharmacol.* 2008;378:503–514.
108. van de Lagemaat R, Timmers CM, Kelder J, van Koppen C, Mosselman S, Hanssen RG. Induction of ovulation by a potent, orally active, low molecular weight agonist (Org 43553) of the luteinizing hormone receptor. *Hum Reprod.* 2009;24:640–648.
109. Newton CL, Anderson RC. Pharmacoperones for misfolded gonadotropin receptors. *Handb Exp Pharmacol.* 2018;245:111–134.
110. Newton CL, Whay AM, McArdle CA, et al. Rescue of expression and signaling of human luteinizing hormone G protein-coupled receptor mutants with an allosterically binding small-molecule agonist. *Proc Natl Acad Sci USA.* 2011;108: 7172–7176.
111. Blanpain C, Lee B, Tackoen M, et al. Multiple nonfunctional alleles of CCR5 are frequent in various human populations. *Blood.* 2000;96:1638–1645.
112. Yeo GS, Lank EJ, Farooqi IS, Keogh J, Challis BG, O'Rahilly S. Mutations in the human melanocortin-4 receptor gene associated with severe familial obesity disrupts receptor function through multiple molecular mechanisms. *Hum Mol Genet.* 2003;12:561–574.
113. Xiang Z, Litherland SA, Sorensen NB, et al. Pharmacological characterization of 40 human melanocortin-4 receptor polymorphisms with the endogenous proopiomelanocortin-derived agonists and the agouti-related protein (AGRP) antagonist. *Biochemistry.* 2006;45:7277–7288.
114. Xiang Z, Pogozheva ID, Sorenson NB, et al. Peptide and small molecules rescue the functional activity and agonist potency of dysfunctional human melanocortin-4 receptor polymorphisms. *Biochemistry.* 2007;46:8273–8287.
115. Xiang Z, Proneth B, Dirain ML, Litherland SA, Haskell-Luevano C. Pharmacological characterization of 30 human melanocortin-4 receptor polymorphisms with the endogenous proopiomelanocortin-derived agonists, synthetic agonists, and the endogenous agouti-related protein antagonist. *Biochemistry.* 2010;49:4583–4600.

116. Zhang FP, Kero J, Huhtaniemi I. The unique exon 10 of the human luteinizing hormone receptor is necessary for expression of the receptor protein at the plasma membrane in the human luteinizing hormone receptor, but deleterious when inserted into the human follicle-stimulating hormone receptor. *Mol Cell Endocrinol.* 1998;142: 165–174.
117. Gromoll J, Eiholzer U, Nieschlag E, Simoni M. Male hypogonadism caused by homozygous deletion of exon 10 of the luteinizing hormone (LH) receptor: differential action of human chorionic gonadotropin and LH. *J Clin Endocrinol Metab.* 2000;85: 2281–2286.
118. Misrahi M, Meduri G, Pissard S, et al. Comparison of immunocytochemical and molecular features with the phenotype in a case of incomplete male pseudohermaphroditism associated with a mutation of the luteinizing hormone receptor. *J Clin Endocrinol Metab.* 1997;82:2159–2165.
119. van Straten NC, Schoonus-Gerritsma GG, van Someren RG, et al. The first orally active low molecular weight agonists for the LH receptor: thienopyr(im)idines with therapeutic potential for ovulation induction. *Chembiochem.* 2002;3:1023–1026.
120. Guo T. Small molecule agonists and antagonists for the LH and FSH receptors. *Expert Opin Ther Pat.* 2005;15:1555–1564.
121. Yanofsky SD, Shen ES, Holden F, et al. Allosteric activation of the follicle-stimulating hormone (FSH) receptor by selective, nonpeptide agonists. *J Biol Chem.* 2006;281: 13226–13233.
122. van Koppen CJ, Verbost PM, van de Lagemaat R, et al. Signaling of an allosteric, nanomolar potent, low molecular weight agonist for the follicle-stimulating hormone receptor. *Biochem Pharmacol.* 2013;85:1162–1170.
123. Lewin AS, Drenser KA, Hauswirth WW, et al. Ribozyme rescue of photoreceptor cells in a transgenic rat model of autosomal dominant retinitis pigmentosa. *Nat Med.* 1998;4:967–971.
124. LaVail MM, Yasumura D, Matthes MT, et al. Ribozyme rescue of photoreceptor cells in P23H transgenic rats: long-term survival and late-stage therapy. *Proc Natl Acad Sci USA.* 2000;97:11488–11493.
125. Murray SF, Jazayeri A, Matthes MT, et al. Allele-specific inhibition of rhodopsin with an antisense oligonucleotide slows photoreceptor cell degeneration. *Invest Ophthalmol Vis Sci.* 2015;56:6362–6375.
126. O'Reilly M, Palfi A, Chadderton N, et al. RNA interference-mediated suppression and replacement of human rhodopsin in vivo. *Am J Hum Genet.* 2007;81:127–135.
127. Chadderton N, Millington-Ward S, Palfi A, et al. Improved retinal function in a mouse model of dominant retinitis pigmentosa following AAV-delivered gene therapy. *Mol Ther.* 2009;17:593–599.
128. Latella MC, Di Salvo MT, Cocchiarella F, et al. In vivo editing of the human mutant *Rhodopsin* gene by electroporation of plasmid-based CRISPR/Cas9 in the mouse retina. *Mol Ther Nucleic Acids.* 2016;5:e389.
129. Burnight ER, Gupta M, Wiley LA, et al. Using CRISPR-Cas9 to generate gene-corrected autologous iPSCs for the treatment of inherited retinal degeneration. *Mol Ther.* 2017;25:1999–2013.
130. Yu W, Mookherjee S, Chaitankar V, et al. Nrl knockdown by AAV-delivered CRISPR/Cas9 prevents retinal degeneration in mice. *Nat Commun.* 2017;8:14716.

Index

Note: Page numbers followed by "*f*" indicate figures, and "*t*" indicate tables.

D

E

F

G

FSC
www.fsc.org
MIX
Papier | Fördert
gute Waldnutzung
FSC® C083411

Kolibri 360 GmbH
Ferdinand-Jühlke-Straße 7
99095 Erfurt, Deutschland
produktsicherheit@kolibri360.de